Practical Models for Technical Communication

Edition 1.0

Shannon Kelley

Practical Models for Technical Communication

ISBN-13: 978-1-943536-95-5

Chemeketa Press
Chemeketa Community College
4000 Lancaster Dr NE
Salem, Oregon 97301
collegepress@chemeketa.edu
chemeketapress.org

For desk copies or ordering inquiries, contact collegepress@chemeketa.edu.

Cover design by Ronald Cox IV
Interior design by Ronald Cox IV, Cassandra Johns, Brice Spreadbury

Acknowledgments appear on page 356 and constitute an extension of the copyright page.

References to website URLs were accurate at the time of writing. Neither the author nor Chemeketa Press is responsible for URLs that have changed or expired since the manuscript was prepared.

This textbook is made possible, in part, by a grant from John and Bobby Clyde to support faculty involvement in textbook affordability efforts.

Land Acknowledgment

Chemeketa Press is located on the land of the Kalapuya, who today are represented by the Confederated Tribes of the Grand Ronde and the Confederated Tribes of the Siletz Indians, whose relationship with this land continues to this day. We offer gratitude for the land itself, for those who have stewarded it for generations, and for the opportunity to study, learn, work, and be in community on this land. We acknowledge that our College's history, like many others, is fundamentally tied to the first colonial developments in the Willamette Valley in Oregon. Finally, we respectfully acknowledge and honor past, present, and future Indigenous students of Chemeketa Community College.

Contents

Introduction

Welcome to the future of technical communication. This book began with feedback from students in technical communication courses. They wanted a book with better visuals, organization, and models. They wanted a book that was easy to navigate and streamlined for their needs. They wanted a useful and practical book with a price tag that didn't make them break out in a cold sweat. At every step of the way, the instructors who contributed to this book had you in mind.

We invite you to approach this textbook as a term-length usability study. In usability testing, the developers of a product evaluate its performance by observing how the target audience use the product. This textbook is designed to provide you with solid tools, useful models, interesting scenarios, and a vocabulary of technical terms that will allow you to communicate effectively as part of a fast-paced, global workforce.

To get the most out of this book, you must first read the pages assigned by your instructor. After you've done that, take a step back and consider how the book, chapter, or section you just read works as an example of technical communication. Like any text, you should approach this book with a curious and critical mind. If one of the book's models or templates helps you understand a concept, let your instructor know. If we fail to follow our own rules, let your instructor know. You are part of our development team now.

Key Features

» **Abstracts**: The abstracts at the beginning of each chapter demonstrate concise, specific writing that sums up the chapter. This pedagogical element reinforces an essential technical skill.

» **Annotated models**: Marginal notes accompany every model to identify places where the document is working or needs work. The notes encourage you to interact critically with the documents and understand specific examples of the chapter's concepts.

» **Bold terms and glossary**: Each chapter begins with a list of key terms that will be introduced. When important terms are first mentioned, they are always in bold and accompanied by a sentence definition. This book has an entire chapter on technical descriptions and definitions, but each bold term can be treated as a mini-lesson in writing effective definitions. The glossary at the end collects all the definitions for easy reference.

» **Case studies**: At the end of each chapter, case studies invite you to think about ways to apply the concepts you're learning. These compact scenarios show how individuals interact with technical communication principles in the workplace and beyond.

» **Checklists**: The checklists at the end of each chapter break down complex tasks into smaller steps. They are designed to be used, so check off each box to track your progress.

» **Extended technical scenarios**: In each chapter, you will follow a character who must create a technical document. As with any good story, each character encounters challenges, and their drafts serve as a starting place to begin seeing yourself as a technical communicator.

» **Figures, tables, and charts**: Whenever possible, this book aims to translate concepts into visuals to support the text.

» **Instructional captions**: The captions are designed to reinforce the visual information provided by the figures. Rather than restating the book's content, the captions provide bite-sized supplemental information about the concept to enhance understanding.

» **"Looking Ahead" sections**: Every chapter opens with a table of contents in miniature. This section offers a quick overview of the chapter to assist in your preparation for class.

» **Marginal notes**: The notes in the margins reveal the connections within the text by pointing to other chapters where a concept is discussed in greater depth.

» **Problem-Solution Framework**: This framework provides a conceptual model to illustrate the goals of technical communication. This model embraces the importance of purpose, audience, and message and adds to it by emphasizing how technical communicators are hired to solve problems.

» **Traffic signals**: Models of technical documents are accompanied by traffic signals that indicate how finished the document is. A red light indicates a document that is at the beginning stages or insufficient. The hope is that you'll stop at these models to examine why they aren't working. Models with a yellow light show a work in progress. Models with a green light are final versions that are ready to go out into the world.

A Note for Instructors

Chemeketa Press would like to build resources that align with this textbook. If you have a classroom activity, student model, assignment, or syllabus that works with this text, please share it with us. We hope to use these resources (with your permission) to create a toolbox for future instructors who use this textbook. Please feel free to contact us at collegepress@chemeketa.edu.

Chapter 1

Technical Communication Fundamentals

Abstract: Welcome to technical communication. Each chapter begins with an abstract that models concise writing and prepares you for the content to come. As you'll learn in this book, technical communication is a journey from problem to solution. The technical communicator must use and arrange clear, concise, precise, and accurate information to create successful, user-friendly documents. Different media, such as images or videos, often help make this information understandable. The key is to know the material's purpose and its audience, called the "user" in this book, and choose the best form for the message. If users can solve their problems or accomplish their goals using content you create, they win, you win, and your employer wins. This chapter also introduces you to the Problem-Solution Framework that will guide your choices as you become a more proficient technical communicator.

Looking Ahead

1. Why Technical Communication Matters
2. The Problem-Solution Framework
3. Purposeful Communication
4. Characteristics of Technical Communication
5. Creating User-Friendly Content

Key Terms

- » accuracy
- » audience
- » clarity
- » client
- » conciseness
- » data
- » demographics
- » end user
- » fact
- » inference
- » judgment
- » medium, media
- » message
- » mode
- » multimedia/multimodal communication
- » precision
- » Problem-Solution Framework
- » purpose
- » scanning
- » skimming
- » technical communication
- » technical document
- » user profile

Why Technical Communication Matters

Have you ever tried to explain to someone how to tie their shoes? It's much harder than it seems. You've been tying your own shoes since kindergarten and now can do it without thinking. In order to show someone else how to do it, however, you have to take what has become automatic and break it down into small steps in a specific order. This process takes mental dexterity. You have to imagine yourself in someone else's place to be able to teach them how to tie their shoes. If you do your job well, you help them build a new skill and prevent them from tripping over untied laces. Shoe tying and technical communication may seem like two very different activities, but we offer this analogy to help you understand your role as a technical communicator.

Technical communicators help audiences solve problems and break complex topics down into simple steps. That's why you're reading this book. In the pages that follow, you will learn how technical communication shares similarities with the shoe-tying situation. The ability to explain with clarity is crucial on the job, whether you work for a government think tank, an engineering firm, or a preschool. What's more, you can apply these skills at any stage in your professional development.

You might be nervous about a course in technical communication because you think the word "technical" implies learning complicated and tedious material. You might have avoided or postponed taking a course in technical communication, assuming that the skills aren't necessary for your chosen field. This textbook aims to show you that technical communication isn't hyper-specialized, impractical, or intimidating. Think about your field. Training to be a professional means you are training how to be a problem solver. Technical communication is about communicating the most direct and effective path toward a solution. Employers and organizations tackle issues, and they need people with advanced communication skills who can translate ideas into plain English.

Figure 1.1. Types of Technical Documents. Technical documents take many forms. The common denominator in technical communication is the creation of content to meet a specific need and produce a desired result for the end user.

Technical Communication Defined

Technical communication involves generating clear, precise, and accurate content about practical information in a field. A technical communicator creates a purposeful message for a specific audience. This can take many forms (figure 1.1).

Although written content is one method used by technical communicators, it's not the only one. You'll notice that technical communication is also about formatting, layout, and visual design, not just words on a page. As a result, this book uses the phrase "technical communication" instead of "technical writing." This textbook introduces you to diverse approaches to technical communication and a range of communication skills that will be useful in any profession.

Technical documents—the content generated by technical communicators—surround you. For example, a bus stop contains specific information relevant to users of public transportation. Riders need to sort through arrival and departure times quickly and efficiently so they don't miss their bus. Effective design organizes the information: the location of bus stops, the route schedule, and connection points with other bus lines. When a document like this fails to do its job, the consequences are real.

Start noticing where and how technical documents intersect with your life. You'll begin to see examples everywhere. Store directories, the

Depending on your field, a **technical document** may have a different name: deliverable, product, report, text, etc.

washing label sewn inside your shirt, heating instructions for a microwave dinner—these are technical documents, too. As simple as these examples sound, they didn't just happen. Someone thought about you, the user, when they designed the mall kiosk to help you get to that out-of-the-way shop that sells pickle-flavored lip balm. The icons on your favorite shirt's label tell you at a glance how to wash it so it lasts longer. Dinner is saved—as well as your taste buds—by instructions that tell you to let the microwaved mashed potatoes sit for five minutes before shoveling them into your mouth.

Technical communicators are a diverse group. Look around and you'll see several majors represented in your classroom. You might have classmates studying computer science, engineering, business, education, medicine, or human services. Technical communicators could be teachers who provide student reports for extra instructional assistance, nurses who write detailed patient summaries to ensure continuity of care during shift changes, or engineers who create product or process schemas.

This book teaches you how to solve technical problems by focusing on the following concepts in your writing:

» The audience's attributes (the "user")
» The purpose of the document
» The message that will resolve the problem

Effective technical communication involves creativity, discipline, and resourcefulness. Wherever you might be headed after this, you are responsible for using the tools described in this book to make someone's life easier and, sometimes, safer.

The Problem-Solution Framework

In an ideal world, technical communicators wouldn't be necessary. Instead, everyone would work through their daily tasks without encountering problems. That ideal world doesn't exist, unfortunately. Individuals often encounter obstacles—complex technical problems—that prevent them from completing tasks. Most users need outside help to move beyond these obstacles.

  This is where a technical communicator comes in. Technical communicators can use the **Problem-Solution Framework** to develop a solution in the form of a technical document. When technical communicators consider purpose and audience, they craft a solution in the form of a message (figure 1.2).

The **Problem-Solution Framework** is explored in more detail in chapters ⑩ and ⑪.

Figure 1.2. The Problem-Solution Framework. This conceptual framework allows the technical communicator to think through a task, including its purpose, audience, and message, before creating a technical document.

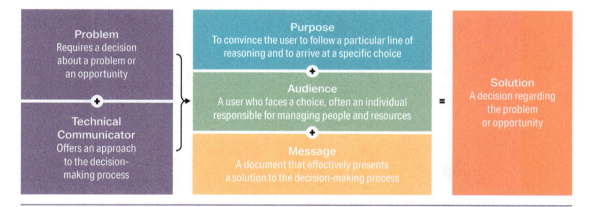

Purpose

Your **purpose** in any technical document is to guide the user to a successful solution. Purpose is the reason for the document's existence and guides all the choices involved in the document's completion. All components of a document need to relate back to the purpose.

Audience

The intended users of the document make up the **audience**. The way you communicate will vary from one document to the next because of differences in audience. You don't guide a professional coder in the same way you do a first-time computer user, for example.

The audience for most technical documents includes a decision-maker. What you communicate—the language and visuals of a document—forms the message.

Message

The document's **message** results from your consideration of the document's purpose and audience. Within this message, you would usually include a call to action for the decision-maker, which could be as simple as "please review and respond by noon tomorrow" or as involved as "this report recommends the replacement of all office chairs with ergonomic models to prevent employee injury." There may be more than one audience for some documents you produce. We refer to those as primary and secondary audiences.

The Framework in Context

One simple way to visualize the Problem-Solution Framework is to think of a three-part approach. The three streams of purpose, audience, and message flow into the solution (figure 1.3).

Figure 1.3. The Problem-Solution Framework: Another View. This three-part approach is another way to think about the Problem-Solution Framework. The document's purpose, audience, and message all work together to create the solution.

Here is another way to think about it. Suppose you need additional storage for your room. You buy a bookshelf from a furniture distributor that sells inexpensive products that customers assemble themselves. If you've ever tried to put together a table, shelf, or bed frame on your own with just the instructions, you know how vulnerable you are. You are at the mercy of the technical communicator.

Let's hope the team that created the instructions kept the Problem-Solution Framework in mind. They know your purpose—to assemble the much-needed shelf. They know about you, the audience—an intelligent but otherwise average person who is not a builder of furniture or even the least bit handy. They know that the instructions (the message) must be simple, clear, and detailed enough to take you step by step through building a shelf that will sit level and hold your collection of vintage records.

You'll encounter the Problem-Solution Framework in future chapters. For now, familiarize yourself with how a technical communicator moves a user from problem to solution.

Purposeful Communication

Now, more than ever, the ways we access and use information change constantly and rapidly. Technical communicators must adapt quickly to the expectations and reading habits of their audiences. For example, when was the last time you read a web page from beginning to end? When you bought your new phone, did you sit down to read the user's manual? How do you find what you need on an online schedule compared with a printed schedule? Think about those questions for a minute. Consider the implications for the design of those documents.

Using versus Reading Documents

In this textbook, technical documents include any mode of communication, whether online or in print, designed to meet a specific need. This means a video, a web page, or a résumé are all considered documents, and it's important to think critically about how you use them.

The concept of "using" documents might be a new one for you. You're likely more familiar with reading them. Why does this distinction matter?

The difference between "using" and "reading" shows how today's audience interacts with documents. For centuries, print media—books, newspapers, and letters—represented the primary way to give and get information. Now, thanks to Google and social media and celebrity cat memes, we live in an age of information overload. We don't interact with content simply as readers. Instead, we look for ways to use attractive content quickly and easily.

Designing for Use

Because of this emphasis on using over reading, technical documents don't focus on text alone, and sometimes not at all. IKEA furniture instructions, for example, present a series of images that show users how to assemble their products. In this way, IKEA communicates a multistep process to their international audience. Customers in Denmark use the same documents as customers in Austria to complete their furniture assembly.

Technical communicators who create documents such as these have power to influence the user's experience, so the stakes are high. But you've got this. Ever since you first learned how to tie your shoes or read, you've been sorting out what's important from what's not. You make snap judgments about the usefulness of a document without thinking about it. Your eye scans for headings, bold text, menus, images, video play buttons, and so forth. You are an information "user."

You are a designer of information, too. Think about the directions you gave to your house for your cousin's graduation party. You explained how to get there turn by turn and included street names. Visitors in the past have complained about being unable to see your house number, so you provided another landmark to guide people to your party. Most arrived as planned and close to on time. Your experience as both a user and designer of information means that you already have significant knowledge about technical communication, although you might not have recognized it as such.

Characteristics of Technical Communication

Technical communication shares characteristics with other forms of communication. Many of the concepts you've studied in other writing or composition courses apply here.

What makes technical communication different, however, is the emphasis on communicating technically complex or practical information. You can see this difference most clearly when you compare technical writing with creative writing (figure 1.4).

Three attributes that distinguish technical communication are its emphasis on multiple modes and media, its focus on the user, and its concern with the needs of the audience. When technical communicators craft their messages, they keep these concerns in mind. We'll take a closer look at these three characteristics of technical writing in the following sections.

Multimodal and Multimedia Content

Technical communicators must weigh the needs of the audience, the technical content, and the form that the content will take. Today's users prefer interactive documents presented in a variety of forms. Lack of choice frustrates users who are accustomed to accessing information in multiple ways.

See Chapter 4 for an extended exploration of multimodal communication.

- » **Modes** are broad categories for how meaning is created, and experienced. Modes come in five forms: linguistic, aural, visual, spatial, and gestural.
- » **Multimedia**, or **medium,** (media is the singular form of medium) refers to the final product that serves as a container for the information.

Figure 1.4. Creative Writing vs. Technical Writing. Examine this comparison of creative writing and technical writing. Consider the context and purpose of each type of writing. The audience's expectations may differ for each.

Ⓐ Creative Writing	Ⓑ Technical Writing
From Jack London's "To Build a Fire"	**How to Build a Fire in the Snow**

	Creative Writing			Technical Writing
Focuses on the main character	[He gathered] dry firewood—sticks and twigs principally, but also larger portions of seasoned branches and fine, dry, last-year's grasses. He threw down several large pieces on top of the snow. This served for a foundation and prevented the young flame from drowning itself in the snow it otherwise would melt. The flame he got by touching a match to a small shred of birch-bark that he took from his pocket. This burned even more readily than paper. Placing it on the foundation, he fed the young flame with wisps of dry grass and with the tiniest dry twigs. . . . Gradually, as the flame grew stronger, he increased the size of the twigs with which he fed it. . . . He worked methodically, even collecting an armful of the larger branches to be used later when the fire gathered strength.	Focuses on the end user (implied "you")	1. Gather at least 3 armfuls of dry sticks, twigs, and grasses. They should range in size from under an inch in diameter (kindling) to 3 inches or larger (logs).
Uses descriptive language		Uses directive language	2. Lay 3 or 4 of the logs on the ground to form a base to protect the early flames from being extinguished by the wet ground.
Uses imagery and metaphor		Uses specific and precise details	3. Lay 2 or 3 handfuls of the kindling on and around the logs. 4. Use a flint or match to strike a flame and hold it to the smallest kindling.
Selects words to create a scene or feeling		Uses active voice to show action	5. Add wood gradually, increasing the size of the sticks as the fire grows in strength until it reaches the desired size.
Varies sentence length and complexity		Uses sentences that are direct and simple	6. Add wood to the fire when it begins to decrease.
Source: Jack London, "To Build a Fire," Lost Face (Mills and Boon, 1919). Public domain.			

Different modes of communication include linguistic (words), aural (sounds), visual (images), spatial (arrangement), and gestural (movement) (figure 1.5). A document with more than one of these modes uses **multimodal communication**. While most YouTube videos involve all five modes, a person with hearing loss interreacts differently with the video's content than a hearing person does. As a result, multimodal communication is shaped by the specific user's experience.

Figure 1.5. Forms of Multimodal Communication. You may not use all these modes all of the time, but you should make conscious choices about which mode or combination of modes best conveys your ideas.

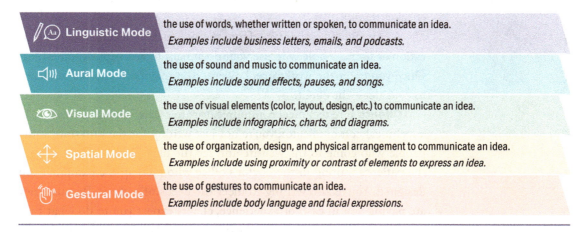

Linguistic Mode	the use of words, whether written or spoken, to communicate an idea. *Examples include business letters, emails, and podcasts.*
Aural Mode	the use of sound and music to communicate an idea. *Examples include sound effects, pauses, and songs.*
Visual Mode	the use of visual elements (color, layout, design, etc.) to communicate an idea. *Examples include infographics, charts, and diagrams.*
Spatial Mode	the use of organization, design, and physical arrangement to communicate an idea. *Examples include using proximity or contrast of elements to express an idea.*
Gestural Mode	the use of gestures to communicate an idea. *Examples include body language and facial expressions.*

Media is the plural form of medium, so **multimedia communication** is when a technical communicator makes use of more than one method of delivery. This textbook is a medium. A PowerPoint presentation is a medium. A podcast is a medium. Think of a web page that has an interactive menu, full-color images, and a video with an introduction to the site's content. That's a good example of multimedia.

User-Focused Content

Consider for a moment the essays, reports, and research papers you've written for other courses. Did you think about your audience? Many students—if they think about their audience at all—focus on what the teacher has stated about the assignment (how many pages, the topic, and other required elements). This is a common view of documents produced for school. The student writes to satisfy the requirements rather than to satisfy a user.

Now, contrast that with text messages you've sent or posts you've put up on social media. Who was the intended audience? For that text message, it was your friend, right? You wrote it for a specific person. And the social media post? You know your followers will see it. Maybe there are even followers you are hoping won't see it (such as your parents or employer). For both messages, you used language and images to capture your audience's attention.

As mentioned earlier, we refer to the audience for technical communication as users, people who use your document to accomplish something. Once you enter the workforce, however, you may have another audience to consider—the client.

Clients hire technical communicators to create content for them. As a technical communicator, you're not just writing for the **end user**—the people who come to the website and see the blog post you've written or the patients waiting in the doctor's office who pick up the pamphlet you designed. You're also writing to meet the needs of a client. Because they are paying you in cash or experience, clients expect you to do what they ask. It's challenging to balance the demands of the client and needs of the end user, but that's what you need to do if you want to keep your job. You must have knowledge of the product, the process, the users, and the client.

Need-Driven Content

To create content for specific users, you must understand their needs. If you miss the mark here, your product will fail no matter how professional it appears. This concept—that you have to know your audience's needs—is essential in technical communication.

How do you get a solid idea of your audience and their needs? First, stop and take a careful look at who your users might be. Get detailed. The more you understand their experiences, the more likely you are to meet their needs. You may need to step outside your comfort zone and get to know more about their culture, demographic, and desires.

Avoid the temptation to say you're writing for a general audience. For example, when writing a persuasive essay for class, your goal is to incite change. Your instructor writes on your paper: "Who is your intended audience?"

"Everyone," you reply. "Everyone should read this paper and change." While this might be true, most successful persuasion targets a specific demographic. To craft an effective message, you should have a purpose and understand your audience. "Everyone" is a tall order. Likewise, if you are trying to start a new club on campus, you need to ask yourself if everyone cares about whether a new club is created. Probably not. Useful technical communication focuses on a specific user. To be successful, you need to be intentional and systematic in understanding what that person wants and needs. One way to do this is to create a user profile.

Creating a User Profile

Each document meets unique needs based on the problem it is solving. The specifics of this problem determine the content and design of the document. You should tailor the presentation of a document to the needs of your particular audience (figure 1.6).

Figure 1.6. Sample User Profile Questions. Use these categories and questions to develop a user profile.

Demographics	User Profile Questions
Ability	Does your audience require special tools, adaptive technologies, or accommodations?
Age	What is your audience's age range? Is your audience multigenerational?
Cultural identity	With what cultures, nationalities, races, and/or ethnicities does your audience identify?
Education	What level of schooling has your audience attained?
Gender identity	How does your audience self-identify their gender?
Income	How much money does your audience make?
Language	What is their native language? Does your audience speak more than one language?
Mode of travel	What kind of car do they drive? Do they use public transportation? Bike? Walk?
Occupation	What kind of job do they have? Are they part-time, full-time, retired, or unemployed?
Technical proficiency	Does your audience have access to technology? What is their level of technological understanding?

Unless you are producing content for a group of people you already know well, you will need to conduct research to understand your audience. A **user profile** collects information about your potential audience assembled through interviews, surveys, reports, or conversations with your client. In other words, a user profile requires you to conduct primary research.

These firsthand accounts can help you determine the user's **demographics**, which are the unique characteristics of your target audience. The more you define your typical users, the more you can anticipate their needs. This includes the ways they might use your content, the potential pitfalls, and the level of detail and explanation you need to include.

A recipe is an example of a need-driven technical document designed for a specific end user who wants to make a specific meal. The content focuses on ingredients, preparation instructions, safety warnings, and alternatives or troubleshooting tips. The design focuses on presenting this content in a way that makes it easy to follow. Compare these two approaches for a soufflé recipe to see how content and design should work together (figure 1.7).

Figure 1.7. Recipe Comparison. Examine these two partial recipe examples. Notice how the content and design differ.

A

A nicely risen soufflé can be impressive, and I think you can do it once you get the hang of it. If you want a cheesy soufflé, you can add Swiss cheese and even some chives to the sauce mix (butter, egg yolks, flour, milk, salt, and pepper) once they're all mixed and cooked together, but before you put it in the oven, which must already be at temperature. Don't fold the whites in all at once. Do it in two batches. Make sure they're stiff, but don't let them get dry. Make sure you prepare the dish so the soufflé doesn't stick. When it sticks, it's almost as frustrating as when it doesn't rise. Timing is also important. You want it to come out of the oven right as you want to serve it.

B

Soufflé
- 2 tbsp butter
- 2 tbsp all-purpose flour
- ½ tsp salt
- Pepper, as desired
- ¾ cup milk
- 4 eggs and 2 egg whites
- ¼ tsp cream of tartar

Preheat oven to 375°F.

Melt butter in a saucepan over low heat and add flour, salt, and pepper.

Cook and stir until smooth. Add milk.

Keep stirring for approximately 20 minutes until smooth and thick.

Creating User-Friendly Content

While many examples of technical content adequately convey ideas, the best forms of technical communication present information in easy-to-understand formats. This does not mean "dumbing down" information for your user. It means meeting the user where they are. The responsibility for creating user-friendly content is yours.

When you write clearly, you show respect for your user's time and attention. You can improve technical documents' usability by making content clear, concise, precise, accurate, and both scannable and skimmable. The following sections will explore these ideas in more detail.

Be Clear

Clarity means creating understandable content through your words, sentences, and organization so users can take action with as little effort as possible. Users come to you, or a document you created, with a goal in mind. They want to accomplish a task, learn something new, or fix a problem. When a document isn't clear, the user may give up in frustration.

To avoid frustrating your users, you need to keep in mind what they already know and what they need to know. Specialized terminology—something you'll encounter frequently in technical communication—can frustrate a general audience. Experts often develop their own language when they talk to each other, and your job as a technical communicator is to translate this expertise into language that non-experts can understand. Just like in the shoe-tying example at the beginning of the chapter, technical communicators must break down new information into smaller, incremental steps that someone else can follow. Notice the difference in usability and clarity between these two examples (figures 1.8a and 1.8b).

This textbook uses streetlights to show the usefulness of each example.

🔴 A red light indicates a problematic example.

🟡 A yellow light indicates the example is close but still lacks effective elements.

🟢 A green light indicates an example with useful elements.

Clear writing also means presenting content in a logical sequence. An organized sentence presents information in the order the user needs it. Sentences written in active voice tend to be clearer because they state directly who is doing what. For instance, take a second look at the final paragraphs in figure 1.8a. The first letter reads, "For your review, an itemized list of labor and materials is attached." Who has provided this list, and what should be done once it is reviewed?

Figure 1.8a. Letter Draft. Your first draft of any technical document is likely to need some revision.

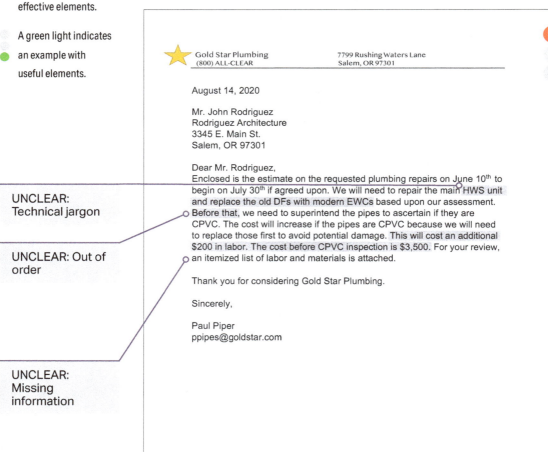

UNCLEAR: Technical jargon

UNCLEAR: Out of order

UNCLEAR: Missing information

Gold Star Plumbing
(800) ALL-CLEAR

7799 Rushing Waters Lane
Salem, OR 97301

August 14, 2020

Mr. John Rodriguez
Rodriguez Architecture
3345 E. Main St.
Salem, OR 97301

Dear Mr. Rodriguez,
Enclosed is the estimate on the requested plumbing repairs on June 10th to begin on July 30th if agreed upon. We will need to repair the main HWS unit and replace the old DFs with modern EWCs based upon our assessment. Before that, we need to superintend the pipes to ascertain if they are CPVC. The cost will increase if the pipes are CPVC because we will need to replace those first to avoid potential damage. This will cost an additional $200 in labor. The cost before CPVC inspection is $3,500. For your review, an itemized list of labor and materials is attached.

Thank you for considering Gold Star Plumbing.

Sincerely,

Paul Piper
ppipes@goldstar.com

A more direct and active rewrite of this sentence can be found in the revised letter in figure 1.8b: "Please review the itemized list of labor and materials I've attached." Now the user knows what to do with the list and who to get in touch with if they have any questions. Clear organization begins at the sentence level and extends throughout the entire document.

Figure 1.8b. Revised Letter. Notice how the second letter takes a step back, considers the audience, and organizes information in a way that is easy to read and understand.

Gold Star Plumbing
(800) ALL-CLEAR

7799 Rushing Waters Lane
Salem, OR 97301

August 14, 2020

Mr. John Rodriguez
Rodriguez Architecture
3345 E. Main St.
Salem, OR 97301

Dear Mr. Rodriguez,

Per your June 10th request, I have completed the estimate for the building's plumbing repairs. We would like to begin on July 30th, with your approval.

We first need to make sure that the pipes are not chlorinated polyvinyl chloride (CPVC). These are known to break easily and could cause significant water damage. They would need to be replaced prior to the other repairs. Once completed, we will repair the main water heater and replace the old drinking fountains with new electrical ones.

The total cost including the pipe inspection is $3,700. If we need to replace the pipes, it will be an additional $500, for a total of $4,200. Please review the itemized list of labor and materials I've attached, and let me know if you have any questions or concerns. Thank you for considering Gold Star Plumbing.

Sincerely,

Paul Piper
ppipes@goldstar.com

CLEAR: Verbal organizers help with sequencing of information.

CLEAR: The technical term remains but the letter now explains what it means.

CLEAR: The numbers add up now. The next step is plainly stated.

Most users won't read every word in a technical document, which is why clear organization is essential. Headings, subheadings, numbers, bullets, and verbal organizers help users track the information and find what they need. Take a look at the two versions of this flyer to see how simple changes can clarify your message, even if you don't change a word (figures 1.9a and 1.9b).

Figure 1.9a. Document Draft. In this first example, no effort has been made to enhance the document's appearance or break up the text.

Verbal organizers help break up the text. This is good.

However, long paragraphs reduce readability. No one really wants to read a paragraph that looks like this.

Tips for Buying a New Car

Buying a new car does not have to be stressful. It can even be a worthwhile experience with a little preparation. Here are a few tips to make car shopping simple.

First, do a bit of homework on your finances before you head to the dealership. Answer the following questions: How much is your old car worth as a trade-in? Visit the Kelley Blue Book website to see how much you can reasonably expect for your vehicle. How much can you afford monthly? Consider how much you are willing and able to finance and for how long. The length of the loan means more interest. Check with your bank or credit union to see what types of loans they have available so that you can compare it to what the dealer's creditors offer. Do you have a down payment? Remember that some dealers have "no money down" offers; others don't. How much are you able to pay for car insurance? Keep in mind that buying a new vehicle will most likely increase your premiums. The more the vehicle is worth, the higher the cost to insure it.

Next, consider your current needs. Answer the following questions: What is the vehicle primary purpose? Think about who and what you will be hauling around—just yourself, kids, other adults, work materials, groceries. How often will you drive it and for how far? Consider gas mileage and whether the car with be a daily driver or weekend fun. What is your style? Make a list of the features you want in a car and how these coincide with your needs. You may like the look of a sporty sedan but need a minivan. What are your options? Do some legwork to find vehicles that meet your needs and offer you a style that you enjoy. Within these, which ones have the best customer and safety ratings? Which ones work within your budget?

Finally, know that the dealer is there to make money—the salesperson is not your friend. If you visit the car lot armed with knowledge, you are less likely to drive away with something you regret. Be patient with the process and shop around until you find a dealer that can meet your needs.

Figure 1.9b. Revised Document. In this revised example, the use of headings, subheadings, bold text, and bullets makes the content easier to read.

Tips for Buying a New Car

Buying a new car does not have to be stressful. It can even be a worthwhile experience with a little preparation. Here are a few tips to make car shopping simple.

Know before You Go – $$$

First, do a bit of homework on your finances before you head to the dealership. Answer the following questions:

- **How much is your old car worth as a trade-in?** Visit the Kelley Blue Book website to see how much you can reasonably expect for your vehicle.
- **How much can you afford monthly?** Consider how much you are willing and able to finance and for how long. The length of the loan means more interest. Check with your bank or credit union to see what types of loans they have available so that you can compare it to what the dealer's creditors offer.
- **Do you have a down payment?** Remember that some dealers have "no money down" offers; others don't.
- **How much are you able to pay for car insurance?** Keep in mind that buying a new vehicle will most likely increase your premiums. The more the vehicle is worth, the higher the cost to insure it.

Know before You Go – Make and Model

Next, consider your current needs. Answer the following questions:

- ✓ **What is the vehicle primary purpose?** Think about who and what you will be hauling around — just yourself, kids, other adults, work materials, groceries.
- ✓ **How often will you drive it and for how far?** Consider gas mileage and whether the car with be a daily driver or weekend fun.
- ✓ **What is your style?** Make a list of the features you want in a car and how these coincide with your needs. You may like the look of a sporty sedan but need a minivan.
- ✓ **What are your options?** Do some legwork to find vehicles that meet your needs and offer you a style that you enjoy. Within these, which ones have the best customer and safety ratings? Which ones work within your budget?

Know before You Go – The Dealer

Finally, know that the dealer is there to make money – the salesperson is not your friend. If you visit the car lot armed with knowledge, you are less likely to drive away with something you regret. Be patient with the process and shop around until you find a dealer that can meet your needs.

Headings grab the user's attention and let them know what the document or section is about.

Subheadings categorize information, increase readability, and allow users to find what they need easily.

Bullets break up the text and single out important items.

Notice the consistent organization. Questions are in bold followed by sentences of similar length and form.

Be Concise

Conciseness means that your document has enough detail and information without unnecessary content. For example, your first draft of an email to your boss might look like this: "The first widget worked better than the second one because it was faster and easier to use." There's nothing wrong with this sentence, but consider your audience. If your boss receives over two hundred emails a day, wouldn't she prefer something more to the point? Your second draft gets you there: "The first widget outperformed the second."

The first time you describe a process or give instructions you might use more words than necessary. That's normal. Make a game out of hunting for every opportunity to trim your sentences, even by a word or two. With time, you can cultivate a habit of rewriting for conciseness.

New writers sometimes over-explain or use more words than necessary to describe a concept. The same can be true of writers who are familiar with a topic—they might struggle to decide what is necessary for the audience. The examples in figure 1.10 translate wordy sentences into more concise versions. Look again, and you'll notice that it's the tiny words that often set up a writer for wordiness: "of," "in," "it," "is," "that," and "to." These are some of the most common words in the English language. You need these words to hold a sentence together. Too many of them, however, can overwhelm the words that carry information and overburden your user.

Technical communication gets to the point. When your content is concise, it means that you have removed redundancy, which is the unnecessary repetition of words, phrases, or ideas. It also means that you've cut out anything unrelated or irrelevant, no matter how clever it seemed when you first wrote it. Think of your sentences like a simple machine. Avoid additional parts that don't help to move it forward.

Be Precise

To make your content precise, focus it like a laser. You should focus on the overall organization of your documents, paragraphs, and sentences, but precision is most evident in the words you choose. When you express your meaning through laser-focused words that meet your audience's needs, you achieve **precision**.

Figure 1.10. Revising for Concision. Study how these sentences move from wordy to concise. Technical communication uses clear, efficient language.

Ⓐ Wordy	Ⓑ Concise
A majority of writers struggle to write in a way that is concise.	Most writers struggle to write concisely.
In many cases, writers are in a position to solve the problem by eliminating fillers.	Often, writers can solve the problem by eliminating fillers.
Another common cause of wordiness is the use of redundant, repetitive, and recurring phrases.	Additionally, redundancy causes wordiness.
Perhaps it could be hedging and the tendency to utilize very unnecessary qualifiers that also contribute to wordiness.	Hedging and unnecessary qualifiers also contribute to wordiness.
The meaning of hedging is when a writer attempts to mitigate a potential loss and safeguard his or her statements with exceptions that allow for unstated contingencies or for a withdrawal from commitment.	Insecure writers hedge to avoid direct statements.* *This wordy example contains so many fillers that a complete rewrite is necessary.

You can make your writing precise by using the ladder of abstraction, a concept developed by English professor S. I. Hayakawa in his book *Language in Thought and Action*. You can think of an actual ladder or an inverted pyramid like the one in figure 1.11 with words on four levels. At the bottom level are words that are specific and concrete. As you move up the ladder, the words become more general and abstract. Precise language often includes words lower on the ladder of abstraction that can be understood through the senses. For example, you know what food is, but you can taste and see a crisp, red Fuji apple.

A thesaurus can also be a great way to find the exact word you need. You can find this reference guide as a book, online tool, or built-in feature of your word-processing program. This resource contains lists of synonyms that can help you expand your vocabulary.

A thesaurus must be used with caution, though. New writers often select random words from the thesaurus to make their writing sound more intelligent. This practice decreases your clarity and precision. Writer Stephen King warns against the use of a thesaurus because "any word you have to

Figure 1.11. From General to Specific. Precision in writing requires words that we can see clearly and recognize. The inverted pyramid illustrates how you can move from general to specific language.

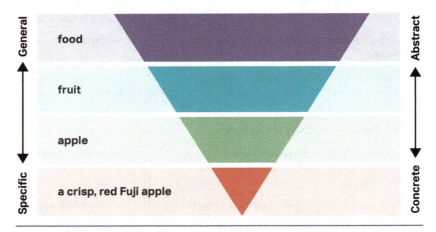

hunt for in a thesaurus is the wrong word." The best word, in his opinion, is the word you already know and the word you know your audience knows.

Consider the different meanings for each of these synonyms for "old":

» Obsolete
» Vintage
» Old-fashioned
» Neolithic
» Classic

If you are talking about a piece of software, calling it "obsolete" would imply that it is no longer in use and has lost its value. Calling it "vintage" means it may have accrued value for collectors and is still being used in some fashion. Either of these words is more precise than the word "old," but they imply different qualities.

While native and nonnative speakers of English know thousands of words, a person's daily use of English is most certainly a fraction of that total. Rather than looking in the thesaurus for fancy new words, search for the familiar word that conveys precisely what you want to say.

Be Accurate

While precision deals with using the most appropriate content for the situation, accuracy refers to using content based on verifiable information. **Accuracy** means that a document is free of errors. Keeping a document accurate requires that you avoid making assumptions or telling yourself that it's "close enough." This means you need to research any gaps in your knowledge before you try to instruct, inform, or persuade someone to do something. As you aim for accuracy, make sure you have a clear distinction among facts, inferences, and judgments.

Facts

Facts are verifiable. When you write a college paper or a researched report, you provide documentation of your sources so your audience (your teacher) can check that your information is factual. In many technical fields, facts are data. **Data** are the bits of information—typically numbers—a communicator uses to support a larger idea.

A single data set needs context and explanation to be factual in a useful sense. For example, try to make meaning out of this data: The National Highway Traffic Safety Administration (NHTSA) states that there were 34,439 fatal crashes in 2016. This single data point is interesting, but it needs more context to be useful. Is that number an increase or decrease from years prior? What is the distribution of these fatal crashes? Are fatalities more common in certain states than others?

Inferences

Inferences are conclusions based on data. When you make an inference from the available facts, you make meaning of the data. You need more than one data set to reach any reasonable conclusions; otherwise you risk making false statements. In many technical fields, it's expensive or dangerous to make recommendations based on insufficient data.

As you saw in the previous example, the 2016 statistic about fatal auto crashes can't support an accurate inference. The fact merely exists. If you add another fact to the first one, you can begin to make connections that lead to a reasonable inference. For example, of the 34,439 fatal crashes in 2016, 10,111 were caused by speeding. What do these facts, side by side, tell you?

Judgments

When you form an opinion based on facts, inferences, and values, you express a **judgment** in the form of a recommended course of action. As a technical communicator, you need to share judgments that are accurate and ethical.

Based on the fatal crash data on speeding, you could make a judgment that "if fewer people drove above legal speed limits, fewer crashes might occur, so people should not speed in their cars." This would be a reasonable and accurate judgment, given this data.

Be Scannable and Skimmable

Today's users expect technical documents to be easily scannable and skimmable. **Scanning** means to look for a specific piece of information in a document. If you need to call your doctor, you might scan the doctor's website for different ways to contact the office. Or you might search for the return policy on a receipt for a sweater that didn't fit right. In both cases, you know the information is available—you just need to find its specific location.

Skimming is different. **Skimming** means to look for the general or main ideas of a document. Readers do this to get a sense of the overall organization or key ideas in a document. Newspaper headlines are a great example: if you skimmed a newspaper and read only the headlines, you would have a general idea of that day's news. You can use this tactic with scholarly articles and studies, which can be long and dense. If you skim the abstract, introduction, and discussion, you can get a sense of what the study covers and whether it applies to your research.

Web pages are often designed to be scanned. Take a moment to notice how your eyes move across this screenshot of a web page in figure 1.12. The visual elements on this page allow your eyes to scan for useful information. Your eye will most likely land first in the top left-hand corner of the page because that is where our eyes have been trained to go ever since we learned how to read. Where does your eye go next? The icons on the right draw the user to the most frequently accessed information on the site. And next? Your eye bounces to the search bar where you can begin looking for materials for your next report.

Figure 1.12. Design for Scannability. Web pages are designed to be scanned. The directional lines show how the structure of the page helps you find what you need.

Consider the way a potential employer reviews a résumé. Headings, subheadings, bullet points, lists, bold, italics, and white space can help direct the employer's eye to important information that will make the résumé stand out. This concept connects basic design with content. Effective design allows the user's eye to move over the page and grasp the major points without sustained effort. It is a new form of literacy. Developing your ability to create clear, precise, accurate, and usable content is why you're here.

See Chapter 3 for more on design and layout and Chapter 6 for more about effective design of job materials.

Case Study

User-Focused Content

This case study is an opportunity for you to put into practice what you've learned. Part of this chapter focuses on the end user and creating user-friendly documents. Look at the following case study to consider what happens when the user's experience is not taken into account:

In her second year at Acme Community College, Maribel began looking at four-year universities. She found a school in Southern California that looked like a good fit. So Maribel scheduled a tour of the campus, purchased a flight into the Orange County airport, and rented a car with Speedy Car Rental.

At Ground Transportation, she saw a sign for "Rental Cars" with an arrow pointing to the left and for "Shuttles" pointing to the right (figure 1.13). The sign was clear. The only problem was that Speedy Car Rental was not listed.

Figure 1.13. Signage Example. Ineffective signage can cause confusion.

She checked her email from Speedy again but found no further instructions. After asking at the Cheap-O-Cars counter, Maribel dragged her suitcase back to the "Shuttles" location and asked for help again.

While waiting, she noticed a bulletin board with several papers pinned to it. One torn sheet of white paper announced that "Speedy Car Rental has arranged to partner with Holiday Hotel shuttle."

Discussion

» How many instances of technical communication can you identify in this scenario?

» What could have been done to make these technical documents more user-friendly?

» What human factors or environmental conditions contributed to the failed modes of communication you identified?

Checklist for Technical Communication

Problem-Solution

- ☐ Have you identified the purpose of your document?
- ☐ Have you established who your audience is by creating a user profile?
- ☐ Have you checked that your message is consistent with your purpose and audience?

Multimodal Communication

- ☐ Have you selected the mode or modes (visual, aural, gestural, spatial, and linguistic) that are most meaningful for your user?
- ☐ Can your user interact with the document, not just read it?
- ☐ Have you determined the best medium (output) for your document according to the user's needs?

Characteristics

- ☐ Is your document clear, understandable, and easy to navigate?
- ☐ Do you use facts that are verifiable and accurate to increase your credibility?
- ☐ Do you avoid unnecessary information that isn't consistent with the document's purpose?

Conclusion

We hope you now recognize some of the ways that you've already been creating and using technical communication. Technical documents need not be complicated or tedious. Equipped with an introductory knowledge of technical communication, you're ready to look in more detail at the specific considerations that effective communicators weigh in all compositions, such as design, collaboration, or research. The next chapter considers ethical communication. In it you'll learn how to keep your technical communication professional, honest, and accessible.

Chapter 2

Technical Communication Ethics

Abstract: The creation of technical content involves choices. As a communicator, you make rhetorical decisions and design choices. You also make choices about what to include, what to exclude, and how to transfer that information to your user. Sometimes these choices have consequences. Experts will at times encounter ethical dilemmas. When needs conflict is when you must think clearly about where you stand. Most professions have a code of ethics, and this chapter introduces you to guiding principles that will help you avoid distorting information, whether intentionally or unintentionally, and ensure that you respect yourself and the work of others.

Looking Ahead

1. Why Ethics Matter
2. Ethics Defined
3. Ethics at Work
4. Distorted or Misleading Information
5. Using Information Ethically

Key Terms

- » ambiguity
- » circumlocution
- » citation
- » copyright
- » Creative Commons
- » ethics
- » euphemism
- » idiom
- » jargon
- » public domain

Why Ethics Matter

Students and professionals alike need to understand the role of ethics in technical communication. Ethical communication requires that you be responsible and reliable in what and how you communicate. Making the right choice isn't always easy—social pressures, stress, and time constraints can tempt people to cut corners. In spite of this, audiences still expect technical communicators to maintain high ethical standards and provide them with the correct tools to make informed choices.

Recent graduate Kamaal now works in the student advising department for a small college. Kamaal spends part of his time creating technical documents to help students learn about career planning. He needs to produce a document that provides students with up-to-date information about the job market. While researching online, he finds exactly the information he's looking for in a report created by another college. Kamaal considers copying the information, presenting it as his own, and submitting the document for approval to his boss. This seems like a great way to save some time. Besides, who's really going to notice?

Kamaal knows that copying information without acknowledgment is considered intellectual theft. It's a convenient solution, and it's unlikely anyone would notice. But what if Kamaal's boss decides to fact-check his data and finds the original document online? Kamaal didn't stay home from countless parties to study for his degree only to throw it all away with an unethical decision. Instead, he finds the original document, uses the information there to build his own document, and provides a reference list using Chicago style, which is what his boss expects.

Kamaal saw an easy way out, but instead he exercised professionalism by doing the extra work and fulfilling the expectations of his boss. An ethical communicator is just that: a person who is professional in all communication. If Kamaal had made the unethical choice and got caught, he could have lost the trust of his boss or his job. Even if he didn't get caught, the choice to plagiarize is not harmless. If the original information was inaccurate or outdated, the end user—in this case, students—could have made serious missteps in their career planning based on what they read in Kamaal's document.

Ethics Defined

Ethics can be described as a system of principles or morals that determine the actions of an individual or a group. Ethical questions are not easily answered. They are rarely yes-no questions. Most of the time, ethical questions ask about what *should* be done and need to account for complex factors. For example, should you report a potential safety issue in the car your team is engineering or wait for road tests so the project comes in on time and within budget? What do you do when your good friend purposefully leaves information out of a progress report to make sure your team looks good to the boss? What might be the consequences for each choice?

Ethical questions provoke conversation, reflection, and debate. Your answers to these questions reveal what you value and how you want to act in the world. Entire courses are devoted to the study of ethics and can usually be found in philosophy departments. This chapter considers the overlap between ethics and professional conduct as it relates to technical communication.

Readers and users of technical documents need to trust that the information provided to them is true and reliable. You are responsible for creating accurate documents. The words you choose matter. The information you include or leave out matters. Personal ethics are just that, personal. But they form the foundation for most conversations about professional ethics. This chapter provides some ways to think and talk about this complex topic.

Your Ethical Stance

Another way to define ethics is to consider your own ethical orientation. Ultimately, ethics relate to the choices you make and how they affect others.

At some point in your life, you have probably explored your morals and principles in depth. Take this opportunity to consider who you are at this point in your life, how you want to contribute to your chosen field, and how you want to show up in the professional world. Figure 2.1 offers some questions that can help you figure out where you stand.

Figure 2.1. Ethical Stance Diagram. This Venn diagram shows how the categories of self, situation, others, and world create overlapping areas of influence that inform responsible decision-making.

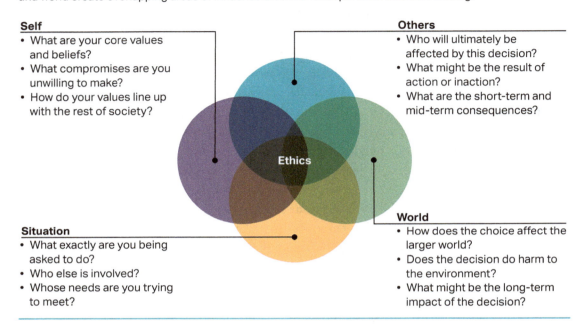

Self
- What are your core values and beliefs?
- What compromises are you unwilling to make?
- How do your values line up with the rest of society?

Others
- Who will ultimately be affected by this decision?
- What might be the result of action or inaction?
- What are the short-term and mid-term consequences?

Situation
- What exactly are you being asked to do?
- Who else is involved?
- Whose needs are you trying to meet?

World
- How does the choice affect the larger world?
- Does the decision do harm to the environment?
- What might be the long-term impact of the decision?

Self

Chances are you were taught a certain world view. Sometimes this aligns with what your parents or guardians had in mind for you, but sometimes it doesn't. At a certain point, it's natural to question those ideologies to see if they are true for you. This is the point where you start to frame your own code of ethics and find the lines you don't want to cross.

Situation

Professional ethics depend on the culture of your workplace, which is where the particulars of the situation come into play. For example, what's permissible in the private sector (nongovernment jobs) may be frowned upon, or even grounds for dismissal, in a government workplace. Looking at the larger context, which includes both the short-term and long-term possibilities, is an important part of ethical decision-making.

Others

Think clearly about the potential impact of your decision on your professional relationships. Behind every technical document is a human being. You have an obligation to the users of documents you create. You are accountable to your employer and client. And you have a responsibility to yourself. How you balance these needs defines your approach to communication. Many of you may create content for global audiences, which requires an even greater awareness of how your choices affect other people.

World

Ethical decisions involve considering not just the impact on people but also the world at large. Technical communicators, specifically those who work in the fields of science and technology, must consider both human and nonhuman factors in making ethical decisions. For example, it would be irresponsible and unethical to ignore scholarly research about environmental issues affecting humans, such as overpopulation, waste disposal, and climate change, to name a few.

Ethics at Work

Professional ethics often involve your legal obligations to your employer. Everything you create while you're being paid is an extension of the company or organization you work for. Each company will have its own standards and expectations, and you should familiarize yourself with them before you begin a new project. These standards can generally be found in your company's code of conduct.

Ethical behavior, including ethical technical communication, involves telling the truth and providing information so that a reasonable audience can make an informed decision. It also means that you act to prevent actual harm, with set criteria for what types and degrees of harm are more serious than others. For example, financial damage to your company outweighs your own frustration or convenience. As a guideline, ask yourself what would happen if your action (or nonaction) became public. If your response could impact your job security or professional reputation—or just cause embarrassment—the action is likely unethical.

But it's not enough to trust your gut. Experts need to confront ethical dilemmas directly by thinking about the possible outcomes and how these outcomes affect various groups of people. How much should you say? To whom? And when? As you look at the examples in this chapter, consider how you might react in a professional situation.

Client and User Conflicts

See Chapter ❶ for more on users and clients.

One of the first places you may encounter an ethical conflict is when your client's needs conflict with your user's needs. It can be tempting to go for the easiest solution—often the solution that pleases your employer (or client) the most. After all, they're the ones paying for the work. A good point to remember is that solutions that appear easy are rarely so. What's convenient for you or your client is not automatically best for everyone involved.

Consider the following example. You've been hired to create a brochure advertising a new prescription drug. The document is a pamphlet that will be picked up and read in physicians' waiting rooms. Your client, a pharmaceutical company, has specific needs. They want the drug to be attractive to a wide audience. They want to sell their product. They've communicated that they want you to focus on the benefits of the drug and minimize the drawbacks. On the other side, users want to know if the drug is safe and if it will help their condition.

Your decision in this situation can be an actual matter of life and death. If you don't give a realistic portrayal of the drug and its interactions, people who encounter your pamphlet may get misinformation that could affect them in serious ways. You're not the only one responsible, of course. There are doctors, FDA regulations on prescription drug ads, and your company's legal team. But your role in creating this information is critical. Being a professional means knowing where you stand and where you draw the line.

Ethics in Action

The Society for Technical Communication (STC) is a professional organization that has identified six ethical principles for its members: legality, honesty, confidentiality, quality, fairness, and professionalism.[1] STC's board of directors adopted these principles in 1998, and they form the standard for professional activity and accountability in the field. You can find the complete code of ethics on STC's website. Figure 2.2 gives an overview of the six principles. Consider how to put these principles into action by taking a look at the case study at the end of this chapter.

Figure 2.2. Code of Ethics. This list of six principles has been adapted from the Society for Technical Communication's ethics statement for professional activities.

Legality	Follow laws and regulations. Fulfill contractual terms.
Honesty	Be honest and accurate; avoid ambiguity. Get permission to use others' work and provide attribution. Use your clients'/employer's time or resources appropriately.
Confidentiality	Get consent to share sensitive information. Obtain releases from clients and employers.
Quality	Do your best work. Negotiate fairly. Meet deadlines.
Fairness	Respect others. Avoid conflicts of interest. Disclose conflicts to those involved.
Professionalism	Provide constructive feedback to colleagues. Request constructive feedback for self-improvement. Get involved in the technical communication community.

Distorted or Misleading Information

In technical communication, visuals are another area where unethical choices can lead to distorted or misleading information. Game designers Leticia and Jason, who you will meet again in later chapters, created this chart to make it appear as though their video game, *Alpacas of Doom!*, greatly reduced the number of alpaca attacks in the U.S. in the months following its release (figure 2.3). At a glance, you can see the line drastically drop immediately after *Alpacas of Doom!* hits store shelves

Look again. What does the chart really say? Leticia and Jason jokingly want the user to believe we are finally safe from vicious alpaca attacks, but they know the data says otherwise. Graphs typically arrange numbers from smaller to larger, so unless someone is reading closely, they might conclude that alpaca attacks decreased after the release of the game. In reality, the number of attacks increased. This visual information takes advantage of the user's expectation for how a graph is supposed to work.

Misrepresenting information is no joke. At best, it will annoy your audience and leave too much interpretation in the hands of the user. At worst, misrepresentation or distortion will lead to faulty decision-making and could result in litigation, injury, or worse.

Take a moment to think about how distorted or misleading information happens. Sometimes miscommunication results from not fully understanding the material. Maybe the writer wants to spare someone's feelings or wants to avoid blame or negative consequences by leaving out details or exaggerating results. A combination of laziness, distraction, lack of experience, competing interests, or any number of human flaws could lead someone to mislead their audience.

Whether or not miscommunication is intentional, misleading or distorted information can lead to serious problems. As a professional, you must set high standards for your work. In this section, we show types of distorted and confusing language so you can recognize these instances and avoid them in your writing. We also provide tips on how to make it right, so you can create technical documents that are accurate, effective, and ethical.

Figure 2.3. Example of Misleading Graph. The graph's design invites a misreading of the data by inverting the numbers on the vertical axis ("Numbers of Attacks") from larger to smaller, rather than the typical arrangement of smaller to larger.

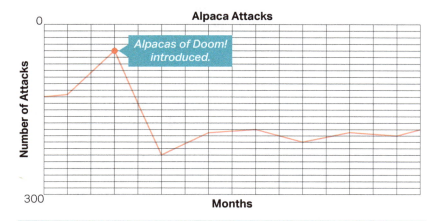

Ambiguity

Ambiguity is when you present multiple meanings at once. Often, ambiguous writing is unintentional. However, intentional ambiguity weakens the effectiveness of technical communication and should be avoided. Intentional ambiguity, or obfuscation, is unethical because the communicator deliberately seeks to mislead readers.

Consider the following statement: "An employee has filed at least one complaint about a supervisor every month for the past year." This statement is ambiguous because it could be understood as multiple employees have filed complaints about a supervisor or one employee has filed complaints about several supervisors. Each scenario differs significantly. Depending on circumstances, complaints from a single employee might be evaluated differently than complaints from multiple employees.

Avoiding ambiguity isn't hard if you determine what's important about the message you're sending. In technical communication, you always lead with important information and follow with relevant additional details. Whether starting a paragraph or writing a sentence, the same principle is true.

Let's say the statement refers to one employee filing complaints against a supervisor and the audience is internal, meaning that no one outside the company will read the report. There is no reason to leave out names in this instance: "Michael Herrera has filed at least one complaint about his supervisor, Dana Samson, every month for the past year." This message becomes important (and damning) evidence about the professional relationship between two people along with a severe failure by the company to deal with the situation.

Relevant information that follows this opening sentence should include details about the nature of the complaints and an explanation of why this issue has continued for a year without being addressed.

Remember that ambiguity is avoided by leading with important information, using active sentences, and making sure your intentions are clear. The next two sections provide examples of specific kinds of ambiguity you might encounter in technical communication.

Euphemism

Euphemisms are words that replace other words to soften the impact of the real definition. Many euphemisms in the English language cluster around taboo subjects, such as death, sex, and bodily functions. For instance, rather than say that someone died, you might say one of the following:

» Harold passed away.
» Fido is in a better place now.

These euphemisms for death attempt to make it sound less frightening. Using more "polite" language to smooth social interactions when dealing with difficult concepts is the most commonly accepted use of euphemism. It provides distance and lessens the blow. Unfortunately, euphemisms can create distance that allows for, or even encourages, irresponsibility. Think about how many ways you can fire employees without taking responsibility:

» The department was downsized.
» Departmental reallocation of resources requires externalizing personnel proportionally.

Bad actors use euphemisms to hide reality, create a legal interpretive space, or make room for active denial later, as these following examples show:

» Enhanced interrogation techniques (torture)
» Collateral damage (civilian death, injury, or property damage during military action)
» Ethnic cleansing (genocide)

In a stand-up routine about euphemism, comedian George Carlin quips: "The more syllables a euphemism has, the further divorced from reality it is." You can use his one-liner to determine when you are writing euphemistically. Is the sentence filling up with syllables? Are your sentences active or passive? Do others know what you mean?

Fix the problems created by euphemism by having an honest conversation with yourself. If you are trying to soften language by using euphemism, explore your intentions. Will the user benefit from this language, or is the benefit only on your end? If the latter, the choice is probably not ethical.

For example, let's look at this 2018 statement from General Motors (GM): "Market conditions require that five North American assembly and propulsion plants will be unallocated product by the end of 2019." This is confusing language that is wide open to a variety of interpretation. An honest, clearer message would read, "Due to the popularity of SUV crossovers, production plants that aren't equipped to build these vehicles will close by the end of 2019."

Bruce Barry, a professor at the Owen Graduate School of Management at Vanderbilt University, suggests GM used "euphemistic language in order to potentially avoid their responsibilities under either the law or their collective bargaining agreements." The word "unallocated" doesn't mean "shut down" or "close," which may save GM from paying severance or offering new jobs and training to their fired workers.[2]

Circumlocution

Circumlocution is the act of burying the message in an overabundance of words. This can happen at the sentence or paragraph level, when the writer overstuffs their document with useless words that tire readers or with language that circles the subject without landing. This tactic can also indicate the communicator has limited understanding of the subject but is trying to appear knowledgeable by increasing word count.

Politicians are well practiced in the use of excessive detail. Here's an example of a politician's response when asked about federal aid for education quoted in the book *Fallacies Arising from Ambiguity* by Douglas Walton:

> I firmly believe that every citizen is entitled to the best possible education. In fact, it is my unalterable conviction that it is the solemn obligation of each generation to endow its youth with the knowledge of the noble achievements of the human species and to make these endowments equally without regard to race, creed, sex, color, or region. I also hold that the burden of providing these rights and privileges should be equitably allotted among those most capable of assuming the burden.[3]

Notice how the speaker avoids committing to an actual answer. Instead, he pumps his audience full of feel-good words that obfuscate the unpopular part of the policy. Try reading this out loud and you'll really hear it. This tactic is effective because the politician not only distracts his audience from the question, but he also gets them nodding along. Let's look at the same passage edited for clarity:

> I believe every citizen is equally entitled to the same high-quality education; it's the responsibility of the present generation to provide this education to the future generations; and we should pay a percentage of tax dollars into education according to our overall wealth.

While arguably not as pretty, the message is clearer: everyone gets the same quality education and wealthier people put more money into the fund that makes equal education possible.

Exclusive Language

Presenting information ethically means considering differences in culture and whether the language in the document is discriminatory. Discriminatory language is often subtle and based on the communicator's assumptions, such as assuming the audience is male or that everyone in the audience is able-bodied. You can easily replace fireman with firefighter. You can also replace words like "walk forward" with "move forward."

In addition to ethical considerations, discriminatory language in technical documents can have serious legal consequences. Responsible technical communicators need to present information in a way that will be understood by a wide range of users, regardless of their background. This means checking documents for language that could be confusing or offensive to readers and using design that accounts for the possibility of visual impairment.

Offensive References

Most cultures find specific words and gestures offensive depending on how they are delivered. In the U.S., the middle finger is typically hostile. In other cultures, giving a thumbs-up is viewed as offensive, which might come as a surprise to people living in the U.S. who associate the thumbs-up with a job well done or an indication that you're ready to proceed. This cultural difference could be pertinent to a technical document.

Graphics and photos in documents that are created in the U.S. sometimes feature a thumbs-up as a means of communicating approval or success. Clearly, "success" isn't how this image would be interpreted by a user from one of the cultures that view the thumbs-up as a rude gesture. For similar reasons, a company based in India won't use a swastika in images meant for an international audience, even though the symbol has religious meaning in Eurasian cultures. They know the symbol is firmly connected with Nazism in Europe and the U.S.—there's no severing that connection.

The meaning of symbols, language, or gestures can change over time. Words that may have had a negative connotation like "sick" (meaning ill) can change to describe something that is cool or amazing. Language evolves so fast, in fact, that by the time this textbook goes to print, the previous example will be out of date.

Technical communicators must be careful to review their content and ensure that it is suitable for users regardless of their cultural background. This requires research, awareness, and responsibility. Some of the work involves self-awareness—often our more casually offensive language is a product of bias or ignorance. For example, assuming only men are part of the audience when women are an estimated 51 percent of the U.S. population or presenting materials that make certain minorities look unsavory or undesirable is not only unethical, it shows poor taste and separation from reality. Those are three traits that don't work well in technical communication fields.

Fix offensive references by researching international cultures for differences from your own culture, reviewing demographic realities, and identifying the biases that may interfere with effective communication. Do a quick internet search using search terms like "business etiquette in (insert country)."

Idioms

Idioms are phrases that have specific cultural meaning and that can't be understood simply by translating the individual words in the phrase. "Learn the ropes" is an example of an idiom. Unless someone has encountered this idiom before, they might not understand that "learning the ropes" means to develop basic skills in an area.

Idioms are particular to a language but also regionally diverse. "Knock on wood" doesn't make sense to English speakers outside the U.S. as a wish for good luck because they're more likely to "touch wood." This is the kind of phrase that you might add without realizing that some users won't understand it.

Fixing idioms isn't rocket science. In fact, it's a piece of cake. The shortcuts we use in casual conversation don't always translate well, especially in situations that require precision. Let's try that again: fixing idioms isn't complicated. In fact, it's an easy task.

Jargon

Jargon is specialized language used by people within a field to communicate concepts anyone in the same profession will understand. For example, consider how two surgeons discuss a procedure with one another. They use specific medical terms to impart important information quickly so the patient survives. If you think about your own professional experiences, you have probably used jargon to communicate with your coworkers—either a shortcut word that everyone at work understands, or a term that only applies to that specific workplace.

Right now, you are learning vocabulary words in your classrooms that are the jargon you're expected to know and understand in your post-college profession. Jargon becomes problematic when technical communicators either forget that a general audience is unfamiliar with the term or use the unknown term against the audience.

When jargon is intentionally used with an audience who won't understand it, without providing clear definitions, the author's motive is questionable at best. Be on the lookout for words designed to halt understanding. A writer who doesn't want the audience to understand is up to no good.

There are many reasons writers might be intentionally unclear: to hide a lack of information, to disguise or downplay the actual situation, to speak directly to people in the know and keep everyone else in the dark, or to persuade the user to do something they want. If you've ever felt pressured into signing a contract you couldn't understand, you have experienced the power of jargon.

On the other hand, many writers are accidentally unclear. Every writer does this, especially in early drafts. Maybe you rushed and didn't review what you'd written. Maybe you thought you were clear but couldn't see the writing from the user's perspective. This is not an ethical problem. It's more an issue of taking your time, reading through your work, and getting second opinions. Still, the outcome is identical to intentional obfuscation: lack of clarity halts understanding and thus action, which is never the goal of ethical technical communicators.

Visual Misrepresentation

See Chapter ❸ for more on visual design.

Images are a common form of information that can be misleading, as you saw earlier with the graph about alpaca attacks. Before you create a graph or insert clip art into a report, think about how your audience will interpret it.

Figure 2.4 is another visual example of information distortion. The person who created this chart may not have intended to distort or misrepresent the information. But, as a professional, you need to provide information in a way that allows your user to draw reasonable conclusions. A student looking at this chart might assume Professor C is a better teacher or an easy grader. But what if the difference is that Professor A is part-time and Professor C is full-time? There's no way to know for sure how this information should be interpreted. Leaving out pertinent information is not just sloppy. It can lead to real consequences.

Other ways to distort information involve the document itself. One of the most common culprits is the three-dimensional pie chart (figure 2.5). Compare Item A with Item C. How different are they in size? It's hard to tell in this chart, isn't it? But in the flat chart, it is clear that Item A is more than twice the size of Item C. Now, look at Items D and B. They're both the same percentage, but the chart's three-dimensionality distorts visual understanding of the ratios. As a result, Item D looks much larger than B. It looks like it is over fifty percent of the graph. The whole idea in a pie chart is to demonstrate how a whole is divided into parts. The three-dimensional graph may look sleek, but it doesn't give a clear representation of the data.

Using Information Ethically

Many of the projects you'll encounter as a student and as a professional require the use of existing material. In fact, any time you perform research for a project, whether it's locating information or including others' points of view, you'll likely need to reference source material that you didn't create. When this happens, you'll need to be aware of the restrictions on using source material. This relates to ethics because how you acknowledge the work of others can have personal and professional consequences.

Figure 2.4. Example of Misleading Chart. Misleading charts create possible ethical dilemmas.

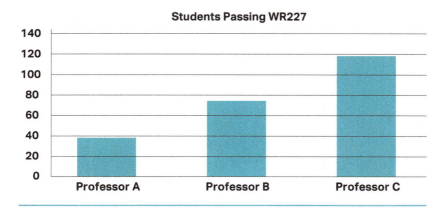

What are some assumptions your user might make based on what they see in this chart?

What could make this chart more useful and reliable?

Figure 2.5. Example of Distorted Information. The three-dimensional version of this pie chart lacks important information to help the user interpret its data.

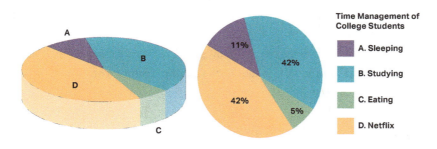

Two major concerns exist for anyone who uses material they didn't create: copyright and citations. **Copyright** is the legal protection that exists for people who own their content. They have legal control over how that content gets used. Any use of that content by someone who isn't the owner is subject to the law, which requires them to get permission from the owner before they use the content. **Citations** are the academic convention that gives credit to the owner of content and also provides an ethical way to use content that may or may not be under copyright.

In academic settings, you don't need permission from the owner to use content because of something called educational fair use, but you always need to use citations to give credit to the content's owner. In professional settings, you need permission from the content's owner to use work that is under copyright.

Copyright Basics

If you want to use content that is owned by someone else, you can verify its owner by locating the copyright symbol (figure 2.6).

The name beside the copyright symbol will tell you who the owner is. For example, this textbook is owned by Chemeketa Community College, the parent organization of Chemeketa Press. The copyright symbol on the copyright page at the front of this book states, "© 2021 by Chemeketa Community College." To use content from this book in a professional setting, you would need to contact Chemeketa Community College and obtain legal permission. It's your responsibility to get permission for content before you use it.

Other common designations for ownership of content are public domain and Creative Commons. Content in the **public domain** can be used without permission, but like asking for permission, it's your responsibility to verify whether the content is free to use or not. The Copyright Office of the U.S. government provides up-to-date information about locating public domain content and verifying its status at www.copyright.gov.

Creative Commons is a nonprofit organization aimed at providing legal and free alternatives to copyright to allow content creators to choose how their work is used or adapted. They have developed a set of licenses that designate what kinds of use are allowed for specific content so that the owner can choose a license that suits their needs. People who opt to use that content are responsible for following the rules of the license the owner chooses. There are four main Creative Commons licenses (figure 2.7).

These licenses can be found in any combination, and their names and symbols are simply listed after the letters "CC" to indicate what the license requires. It is your responsibility to identify which license the content is under and to follow the requirements of that license.

Figure 2.6. Copyright Symbol. A work does not need to have a copyright symbol or be registered with the U.S. Copyright Office to be protected under copyright. Work that you create, called an "original expression" in copyright law, is automatically protected.

Figure 2.7. Four Types of Creative Commons Licenses. Creative Commons licenses provide additional means for content producers to grant permission and receive credit for their work.

BY: Attribution	This license requires that users give credit to the owner by name and, in most cases, provide a URL that leads back to the original content.
SA: ShareAlike	This license requires that any content which is produced using the owner's content is also licensed under an identical Creative Commons license.
NC: NonCommercial	This license requires that the primary purpose of any content which is produced using the owner's content is not to generate commercial profit.
ND: NoDerivatives	This license requires that any content which is produced using the owner's content is identical to the original, and that no changes, adaptations, or "remixes" can be made.

Citations

In an academic setting, any content from outside source material must be acknowledged in the form of a source citation. If you've taken previous writing or communications courses in college, you should have experience with this. There are many citation styles for academic work, and each style has its own set of specific rules and requirements for what information is included in the citation and how it is formatted in your work.

See Chapter 5 for more on citing sources.

What Should You Cite?

The short answer is you should cite everything you use in your research that you didn't generate. This is true whether you're quoting directly or paraphrasing from the source. In technical communication, the chances that you need to quote any source directly are slim to none. You are much more likely to use data from studies, fieldwork, or meta-analyses and reserve direct quotes for instances in which a source states an idea in the best possible way. When you cite sources in your technical document, you are providing the user an opportunity to quickly find and review the original research.

How Should You Cite?

Academic citation standards are field-specific, which is why you will learn several citation styles in college. There are similarities among the styles, but the differences are what make the citation style useful for experts in that field. As a student, it's a good idea to check with your professors to

Find **citation guidance** at owl.purdue.edu.

see which style they expect you to use. Citation styles are updated regularly, so you should also invest in an updated reference guide that provides templates and models to follow for all source types. Most colleges have a preferred reference guide. There are also free online guides like Purdue Online Writing Lab (OWL).

Modern Language Association (MLA) is the format used for literature, languages, and humanities and is one of the two main formats you will use most in college. MLA uses an author/page number in-text citation style with a Works Cited page. This means the information you use from an outside source is identified in the text with the author's name and the page number of the original source enclosed in parentheses. The information included in parentheses connects to a list of sources at the end of your essay.

Here are a few things to keep in mind about MLA style:

» You are unlikely to use MLA in the workplace because it is a scholarly format.
» You are most likely to use MLA in your English and composition courses because it works well for textual analysis.
» MLA is currently in its 9th edition.

American Psychological Association (APA) is used in behavioral and social sciences and is one of the other main academic styles you can expect to use in college. APA uses the author/date citation style with a reference list. Information from an outside source is followed by parentheses containing the author's name and the date of publication. The dates are an important part of the APA citation style because the fields that use this style want the reader to be aware of how recent the research is in the paper.

Here are a few things to keep in mind about APA style:

» You may use the APA format in sociology, anthropology, psychology, and some other science courses.
» You will see APA format in science writing and peer-reviewed scientific articles.
» You will encounter APA format in any number of fields, including nursing programs and exercise/fitness professions.
» APA is currently in its 7th edition.

Chicago Manual of Style/Turabian (CMS) is used in the publishing industry for formatting books and citing references. Turabian Style refers to the student version of the CMS style that uses either the author/date/page number format or the (more common) footnote style and a bibliography. This style is used by a wide range of publishers and commonly used by business, history, and fine arts. This textbook utilizes Chicago style.

Here are a few things to keep in mind about Chicago/Turabian style:

» You will see CMS format used in fine arts, business, and history publications.
» You will likely use Turabian style in business and history courses.
» You will encounter CMS format in most textbooks, such as this one.
» Chicago style is currently in its 17th edition.

Institute for Electrical and Electronics Engineers (IEEE) sets research and citation standards for electrical, electronics, computer science, and computer programming fields. IEEE uses a bracketed number and corresponding reference list set in numerical order.

Here are a few things to keep in mind about IEEE style:

» You are most likely to use IEEE in computer science and programming fields or in electrical engineering.
» The place most people encounter IEEE is Wikipedia.

What does citing your sources have to do with ethics? It may seem like a flaming hoop you need to jump through in academic courses, but this requirement of keeping track of your sources will benefit you in the professional world.

First and foremost, technical communication values accuracy and precision. Your sources of information must be carefully chosen to provide your audience with the best and most recent information. In-text references, followed by a longer description of each source, allow your audience to find and verify your sources. Second, documentation of sources shows your integrity—in other words, it shows that you are responsible with information and meticulous with your recordkeeping. Third, citations demonstrate confidence in your conclusions and reassure the audience that your recommendations are based on data. Finally, citing sources gives credit where credit is due and shows the reader your professional standards.

Case Study

Ethics for Technical Communicators

This case study is an opportunity for you to put into practice what you've learned. Part of this chapter focuses on the six ethical principles outlined by the Society for Technical Communications. Look at the following case study to to begin thinking about the kind of technical communicator you want to be and how you want to make decisions when the answer is not clear:

> See "Ethics in Action" on page 35.

Let's take a closer look at ethical principles in context by examining how freelance technical communicator Julia navigates multiple professional environments and demands. As a freelancer, she must determine fair prices for contract work, pay attention to legal and contractual obligations, recognize when to keep information confidential, and weigh her personal values when encountering conflicts of interest.

Consider how Julia applies the six principles of legality, honesty, confidentiality, quality, fairness, and professionalism to her contract work, and then answer the questions that follow.

Legality.

» Julia's contract job states she must produce or revise three pamphlets using data from a marketing research company. Legally, she can't use copyrighted material—this must be original work

Honesty

» Julia is tasked with promoting the field of robotics. Her boss wants pamphlets for interested students. Julia researches the career to avoid misinforming students, estimates costs to produce the pamphlets (including work hours), and seeks permission to use attractive photographs owned by a third party before making the proposal to her boss.

Confidentiality

» Julia contracts with medical companies. A client suggests Julia ask for patient information so she can research patient reactions to a drug she has been asked to market. Julia knows she can't have this conversation with the doctor's office because it's a violation of the Health Insurance Portability and Accountability Act of 1996 (HIPAA).

Case Study, continued

Quality

» A client contacts Julia about designing promotional materials for a local gym. Julia estimates six hours of initial work and three to five revision hours. She has a design background but limited experience, so she researches fair prices and finds the average rate for experienced designers is about $75 an hour, while someone with her experience charges around $30. Julia provides an estimate of $330 (using the high end of hours she thinks the project will take). She provides documentation on how she reached her estimate.

Fairness

» Julia's work impressed gym customers, resulting in new clients. She's approached by a business that supports policies at odds with the volunteer organization where she helps at-risk youth find education and job opportunities. Although she might make some extra money, she politely declines the work, citing the conflict between the business and her volunteer work.

Professionalism

» Julia makes sure to get an early draft or two to her employer or her clients for feedback because she knows it shows professionalism, and the early feedback she receives will save everyone time and money.

Discussion

» Think about the principles and your career or particular field of study. Can you think of an example for each of these principles in your field?
» Can you explain to someone why each of these principles is important in your field?
» How would you adapt these principles to apply to your field of study? What would you add?

Checklist for Ethical Communication

Citations

☐ Have you acknowledged your collaborators?
☐ Have you used the citation style your field of study requires (MLA, APA, other)?
☐ Are your sources clearly and accurately cited?

Equitable Language

☐ Did you avoid exclusive and offensive language?
☐ Does your text adhere to Plain Language rules?
☐ What checkpoints have you enlisted to locate and change ambiguous use of words?

See "Plain Language Guidelines" (figure 9.5) on page 242.

Visual Details

☐ Have you used the most suitable type of visual (pie chart, table, or diagram)?
☐ Did you give credit to your sources and cite all visuals?
☐ Did you explain to the user what information to use from the visual?

Conclusion

As you move forward in your study and practice of technical communication, remember the principles covered in this chapter. Like Kamaal or Julia, you will probably feel pressured to take an ethical shortcut, sooner or later. Recognize that the temporary advantages are not worth the long-term damage to your reputation. Follow the requirements of employers, as well as legal standards. Communicate clearly by avoiding approaches that confuse or exclude. Don't manipulate ideas to make your job easier. Get permission to use information that is not yours and give credit for that information.

Ethical considerations should not be an afterthought. Instead, make ethical communication the basis of all your work. Behaving ethically means technical communicators put the needs of clients and users above their own. Being an ethical communicator doesn't need to be complicated, but it does require a professional frame of mind. Your relationship with clients and users is based on trust. Honor that trust by preparing your communication as clearly, precisely, accurately, and honestly as you can.

Notes

1. "Ethical Principles," Society for Technical Communication, Adopted by the STC Board of Directors 1998, https://www.stc.org/about-stc/ethical-principles/.

2. "Worst Corporate Euphemism Ever? GM's 'Unallocated' Factories a Contender," *CBSNews*, November 27, 2018, https://www.cbsnews.com/news /worst-corporate-euphemism-ever-gms-unallocated-factories-a-contender/.

3. Douglas Walton, *Fallacies Arising from Ambiguity*, Applied Logic Series (Springer: 1996), 175.

Chapter 3

Layout and Design

Abstract: Technical communication requires that you understand how documents are constructed. At its core, document design is about organization. Words and images must work together to create a technical document. Whether you are a writer or a designer, your job involves gathering information into a meaningful format. This chapter explores design principles that produce accessible and inviting content to guide a user through a document. Ultimately, technical communication involves collaboration, not just among people who create the different parts of a technical document but also among the elements of document design, organizational design, and visual design.

Looking Ahead

1. Why Layout and Design Matters

2. Principles of Document Design

3. Design for Readability

4. Design for Emphasis

5. Design for Organization

6. Types of Visuals

Key Terms

- » alignment (center, right, left, justified)
- » chart
- » chronological organization
- » contrast
- » design
- » diagram
- » document design
- » font (serif, sans serif)
- » graph (bar, line)
- » grid
- » heading
- » illustration
- » layout
- » parallelism
- » problem-solution organization
- » proximity
- » sequential organization
- » subheading
- » table
- » white space

Why Layout and Design Matters

More than ever before, how information is presented affects every aspect of our lives. Principles of layout and design determine how you navigate a website, how you use the self-checkout at a grocery store, and how you interpret the nutritional facts on the packaging of your favorite burrito. Even if you don't think about the individual parts of product communication, you're still influenced by the overall effectiveness of its layout and design. When a design strategy is successful, you likely won't notice. But when these parts lack a clear message or purpose—such as when a user manual doesn't clearly define the functions of your new electronic device—you might find yourself irritated and unable to complete a task.

A poorly designed document can become a source of frustration. Users must deal not only with the difficulty of the problem they are attempting to solve, but also with the challenge of an inadequate technical document. As a communicator, you need to consider how your organizational choices impact users. Thinking directly about document design will directly influence the success of your communication and significantly improve your users' experience.

Layout and Design Defined

Layout describes the intentional placement of visual elements within a document. In general, **design** is the structure or plan of anything that is used. In particular, design is what guides a person's interaction with an object and its features. The two terms are often linked as if they are one thing: layout-and-design. However, you can think of design as the big picture and layout as the details.

Consider how you use applications, or apps, on a phone or computer. Design is what makes the difference between an app that's easy to use and one that isn't. Layout can be seen in the designer's choice to place the navigation on the top of the screen instead of the left. If you can easily find what you're looking for, you can assume the layout and overall design of the app is effective. Simply put, design determines the quality of your experience.

Effective design in technical communication shares many qualities with a well-designed app: it should be easy to use, purposeful, and tailored to the user's needs. These qualities don't just happen. They result from the thoughtful application of design principles and layout elements that we'll explore in this chapter.

Design isn't just about how a document looks. Design refers to the underlying structure of your document and how you group and organize information to make it accessible to your user. While visuals are a part of that, other elements are equally important, such as balance, grouping, consistency, and contrast. This chapter takes a look at how these elements relate to readability, emphasis, and organization in technical communication.

Design and Structure

As discussed in previous chapters, designing for users instead of readers affects the choices you make in the design process of technical documents. With layout and design, just as with other elements, your goal is to provide users a solution to their problems. Your design choices should allow the user to navigate the document quickly and easily to find a solution. Layout and design factor into this solution.

Effective design is intentional. Haphazard or poor design choices create confusion. Consider the designer who chooses a font with a whimsical appearance, even though the subject matter is serious. In this case, the design might frustrate or mislead the user rather than clarify the purpose. Poor design choices actually make finding a solution more difficult for the user.

Effective layout allows a user to glance at a document and understand its purpose. In the case of instructions, for example, you should begin with the user's need to complete a specific task. Design choices should support that goal. Consider how the order and arrangement can move users through each step of the process.

Figure 3.1. Signage – Ineffective (A). This sign's design sends mixed messages.

Figure 3.2. Signage – Effective (B). This sign clearly communicates what it wants.

User expectations should be considered when making design choices. Be familiar with common design schemes, especially as they relate to the type of document you're creating. For example, the color red, in the U.S., is associated with warnings or corrections in technical documents (figure 3.1). If you choose a red font just because you like the color, you might confuse users or in some cases offend them. Figure 3.2 uses a red font effectively to tell users what not to do.

Design at Work

The following scenario shows one example of how design is used in professional settings. Kenji works for Step-by-Step Advertising and PR. His client, Tavent Inc., wants to promote a new industry conference in India. Tavent has asked Kenji's team to design a brochure in both digital and print formats to promote the new conference. The brochures need to be compelling and easy to understand. The layout needs to be organized and user-friendly.

Even the simplest trifold brochure requires the technical communicator to juggle multiple design elements. Kenji needs to make the information accessible and appealing through a combination of words and images.

He needs to choose a layout to organize relevant information (figure 3.3). In this case, relevant information includes a description of the conference, the date, the schedule of events, registration details, and cost. People want to know when the new conference is happening, what it's all about, and if it's worth their time.

Figure 3.3. Brochure Template. This brochure template shows Kenji's initial document design with areas in gray reserved for images and areas in green for text.

Thinking back to the relationship that design and readability share, Kenji's team already has their purpose: to promote a new conference. And they know who the audience is: engineers. Knowing their purpose and audience helps the team identify the message or type of content needed: date, events, costs, etc. The team's next step includes choosing a page layout and design elements that will invite and guide the user through the document.

Principles of Document Design

Document design brings together elements of page layout so users can solve a problem. Technical communication design differs from what you may have learned in academic courses about page layout. In most college courses, your assignments require a double-spaced page with 1-inch margins in 12-point Times New Roman font with indented paragraphs.

However, in technical communication, your user's needs come first. Some documents' design will be more straightforward, as in a business letter. Others will have more design decisions. The key to effective design is simplicity and consistency. A document that is unnecessarily complex makes it harder for users to get what they need.

Design is the deliberate organization of text and images on a page. A children's book is full of colorful images, short blocks of text, and larger font sizes. The book designer, in cooperation with the writer and publisher, chooses the book's design to ensure the audience (children) can easily understand and use the material. As you can see in figure 3.4, the many elements of document design, visual design, and organizational design must work together to create a functional document.

Figure 3.4. Elements of Visual Communication. Effective visual communication involves the overlapping areas of document design, organizational design, and visual design.

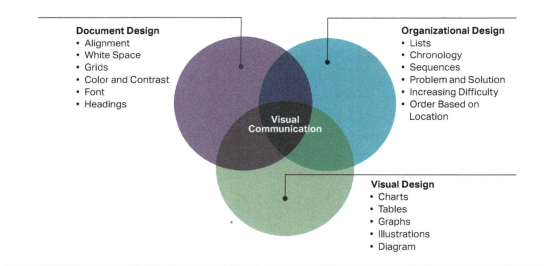

Document Design
- Alignment
- White Space
- Grids
- Color and Contrast
- Font
- Headings

Organizational Design
- Lists
- Chronology
- Sequences
- Problem and Solution
- Increasing Difficulty
- Order Based on Location

Visual Communication

Visual Design
- Charts
- Tables
- Graphs
- Illustrations
- Diagram

Design is equally important in the forms you fill out, such as the ones found at the doctor's office. These black-and-white forms contain standard options: multiple choice, fill in the blank, or basic yes-no questions. The forms might ask you to write out your name, starting with your last name first. Official forms often use this layout because it makes it easier for the doctor and office manager to keep track of and file multiple forms filled with patient health information. The form leads with the critical information, your last name, to which all your medical records are attached. Take a look at the medical form in figure 3.5. The form appears straightforward, but all the elements are intentionally placed.

Figure 3.5. Design Choices in Medical Form. Even a simple form has "hidden" design elements that influence the user's experience.

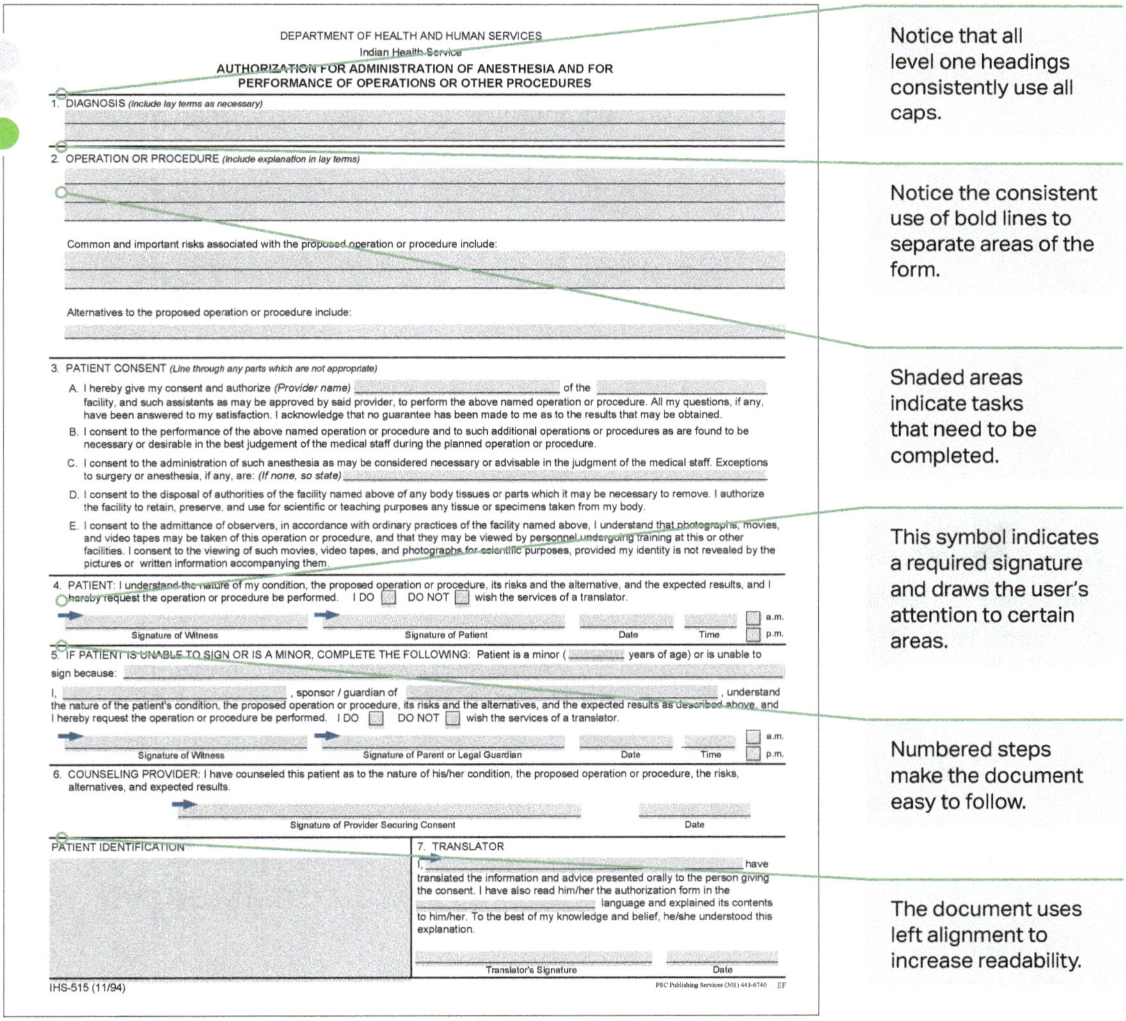

DEPARTMENT OF HEALTH AND HUMAN SERVICES
Indian Health Service
AUTHORIZATION FOR ADMINISTRATION OF ANESTHESIA AND FOR PERFORMANCE OF OPERATIONS OR OTHER PROCEDURES

1. DIAGNOSIS *(include lay terms as necessary)*

2. OPERATION OR PROCEDURE *(include explanation in lay terms)*

Common and important risks associated with the proposed operation or procedure include:

Alternatives to the proposed operation or procedure include:

3. PATIENT CONSENT *(Line through any parts which are not appropriate)*

A. I hereby give my consent and authorize *(Provider name)* ______ of the ______ facility, and such assistants as may be approved by said provider, to perform the above named operation or procedure. All my questions, if any, have been answered to my satisfaction. I acknowledge that no guarantee has been made to me as to the results that may be obtained.

B. I consent to the performance of the above named operation or procedure and to such additional operations or procedures as are found to be necessary or desirable in the best judgement of the medical staff during the planned operation or procedure.

C. I consent to the administration of such anesthesia as may be considered necessary or advisable in the judgment of the medical staff. Exceptions to surgery or anesthesia, if any, are: *(If none, so state)* ______

D. I consent to the disposal of authorities of the facility named above of any body tissues or parts which it may be necessary to remove. I authorize the facility to retain, preserve, and use for scientific or teaching purposes any tissue or specimens taken from my body.

E. I consent to the admittance of observers, in accordance with ordinary practices of the facility named above, I understand that photographs, movies, and video tapes may be taken of this operation or procedure, and that they may be viewed by personnel undergoing training at this or other facilities. I consent to the viewing of such movies, video tapes, and photographs for scientific purposes, provided my identity is not revealed by the pictures or written information accompanying them.

4. PATIENT: I understand the nature of my condition, the proposed operation or procedure, its risks and the alternative, and the expected results, and I hereby request the operation or procedure be performed. I DO ☐ DO NOT ☐ wish the services of a translator.

Signature of Witness | Signature of Patient | Date | Time ☐ a.m. ☐ p.m.

5. IF PATIENT IS UNABLE TO SIGN OR IS A MINOR, COMPLETE THE FOLLOWING: Patient is a minor (______ years of age) or is unable to sign because: ______

I, ______, sponsor / guardian of ______, understand the nature of the patient's condition, the proposed operation or procedure, its risks and the alternatives, and the expected results as described above, and I hereby request the operation or procedure be performed. I DO ☐ DO NOT ☐ wish the services of a translator.

Signature of Witness | Signature of Parent or Legal Guardian | Date | Time ☐ a.m. ☐ p.m.

6. COUNSELING PROVIDER: I have counseled this patient as to the nature of his/her condition, the proposed operation or procedure, the risks, alternatives, and expected results.

Signature of Provider Securing Consent | Date

PATIENT IDENTIFICATION

7. TRANSLATOR

I, ______ have translated the information and advice presented orally to the person giving the consent. I have also read him/her the authorization form in the ______ language and explained its contents to him/her. To the best of my knowledge and belief, he/she understood this explanation.

Translator's Signature | Date

IHS-515 (11/94) PSC Publishing Services (301) 443-6740 EF

Notice that all level one headings consistently use all caps.
Notice the consistent use of bold lines to separate areas of the form.
Shaded areas indicate tasks that need to be completed.
This symbol indicates a required signature and draws the user's attention to certain areas.
Numbered steps make the document easy to follow.
The document uses left alignment to increase readability.

Design choices are not limited to printed documents. Website designers also make choices that help users find information. If the purpose of a website is to sell you remote control cars, for example, the designer might place an image of the car on the left-hand side of the page where our eyes naturally land. The designer might then orient the image of the car so that it points to the purchase button, which is set off by white space on the right-hand side of the page. Every design choice is made to drive the user toward the site's main purpose: a sale.

Principles of design should inform your decisions as you create documents. As you read, you'll notice that some elements show up multiple times under different principles. In the following sections, we look at how the general design principles of readability, emphasis, organization, and visual accessibility make for a more pleasant user experience.

Design for Readability

Users want documents that are easy to read and understand. Effective design elements, including font, color, contrast, alignment, and grids, can increase readability. However, as you can see in figure 3.6, too many design elements can distract from the message.

Figure 3.6. Basic Principles of Design. Fonts can add an element of playfulness to your document, but they may also be less readable and harder for the eye to track. Simple, universal fonts are preferable in technical communication.

THE ESSENCE OF
GOOD DESIGN
IS SIMPLICITY AND
CONSISTENCY.

The essence of
good design
is simplicity
and consistency.

Take a look at figure 3.7 for an example of how Kenji uses design elements to increase readability in his brochure. Building on the brochure template he chose in figure 3.3, Kenji is starting to think about how users will encounter information about the conference and the order and placement of this information.

Font

A font is a set of characters, including letters and numbers, that all have a similar appearance. Font choices can make a document easier to read, add emphasis, or draw the eye to certain parts of a document. Fonts come in all styles, from simple to ornate and from traditional to modern. All fonts fall into one of two categories: serif or sans serif (figure 3.8).

Figure 3.7. Visual Elements for Readability. Kenji's project is coming along. Notice how he uses a combination of visual elements to increase the readability of his conference brochure.

Contrasting colors call attention to important elements.

Large fonts identify titles and create distinct sections.

Grids guide the organization and align text across all three panels.

Figure 3.8. Font Styles. Serif fonts are frequently used in newspapers and books. Sans serif fonts are often used for websites and headings.

Serif — Adobe Caslon Pro

Sans Serif — Helvetica

Serif fonts are more decorative. You can identify a serif font by the additional stroke at the top or bottom of most letters. This textbook uses a serif font for the main text. Historically, serif fonts were the preferred font in newspapers, books, and other printed material. Serif fonts are used for printed text because the small decorative flourishes guide the eye from letter to letter, allowing for faster reading. The most common serif font is Times New Roman, which you've probably used in your college writing classes.

Sans serif fonts, on the other hand, do not have these ornamental pieces (serifs). Sans serif fonts have a cleaner appearance. The no-nonsense fonts of Helvetica and Arial are the most common sans serif fonts. Electronic reading and increased screen time have made sans serif fonts more popular because they are easier to read on different screen sizes and at different resolutions. Because of their simplicity, sans serif fonts are common in road signs, cautions, and instructions.

Font size and style draw attention to changes in a document. For example, Kenji knows that a heading's size allows users to rank the relative importance of the information. Kenji should be deliberate in his design choices involving fonts. Should the font be black or another color? How big or small should it be?

Some people have strong opinions about fonts, such as Comic Sans, Jokerman, Papyrus, or any font that has been overused or used without a lot of thought. For technical documents, readability and simplicity should guide your choice. A common design rule advises against using more than three fonts on a single page.

Color and Contrast

Color and contrast guide users through a text and draw attention to particular areas. However, if used improperly, these same elements can distract users (figure 3.9). As with fonts, you should avoid using more than three different colors in your design. Simplicity and consistency are key.

To make his brochure more inviting, Kenji uses Tavent's company colors. He asked for a style guide from the company so he can be sure to use the company's signature color scheme. This will save the client time because they won't need to edit as much before they incorporate the brochure into their larger proposal.

Color has different associations within different cultures. For example, in Western cultures such as North America and Europe, yellow represents warmth or caution depending on the context and is used in transportation systems for school buses and traffic signals. Yellow, however, represents death and mourning in some Latin American cultures. If Kenji were to create a brochure for a conference in Rio de Janeiro, for instance, he should probably avoid using yellow as a dominant color in his design.

Figure 3.9. Problems with Color. Color can enhance a technical document, but if it is the only means of representing information, it can impact readability. This is especially problematic for people with impaired color vision.

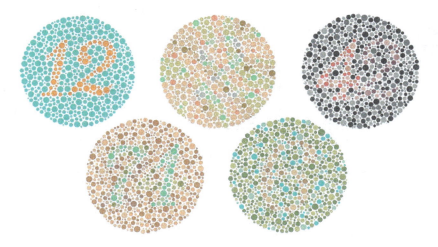

Alignment

Alignment is one of the most basic and important features of document design. You might be so familiar with it that you take it for granted. **Alignment** means that every element has some visual connection with every other element on the page. Figure 3.10 shows the four basic types of alignment.

In **left alignment**, the text lines up along the left margin. This alignment has a ragged edge on the right margin. You may recognize this alignment from academic papers you have written. Kenji used this standard alignment on the market research report he submitted to Tavent because it allowed his client to easily read the multipage report.

In **center alignment**, the text is anchored along the center line of your document. The result is text that creates a mirror effect from one side to the other. Center alignment can be a great way to make sure your white space is balanced. Kenji will center the name of the conference on the front panel of his brochure to make it stand out.

In **right alignment**, the text lines up along the right margin. This alignment is typically used to draw attention to a block of information. In the brochure's design, Kenji might align the date and time of the conference to the right-hand margin to draw attention to those details and to set them apart from the rest of the text.

Figure 3.10. Types of Alignment. In the first three examples (A, B, and C), the content is either anchored to the left-hand margin, the center line, or the right-hand margin. With justified alignment (D), the content is distributed evenly between the left and right margins.

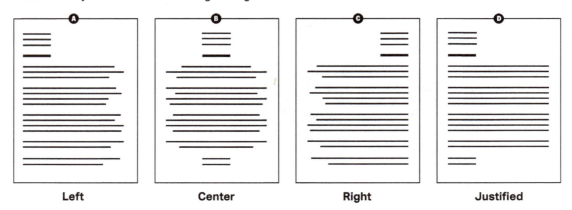

| Left | Center | Right | Justified |

In **justified alignment**, the text aligns along both left and right margins. You see justified text most often in book-length documents (like this textbook) and professional publications, such as journal articles and booklets. If Kenji includes descriptions for the breakout sessions in the brochure, justification will create clean lines down both sides of the fold out.

Ineffective alignment, as shown in figure 3.11a, is the result of not considering how the placement of text and images affects the user's experience. In figure 3.11b, you can see how effective alignment connects the user's eye with different elements on the page.

So, which alignment is the best choice? Think first about how the document will be used. Is it in print or online? Users read these types of documents in different ways. Does the document include long passages?

Alignment allows users to scan a document more efficiently. As in most aspects of technical communication, keep your user's needs in mind. Don't make your choice of alignment simply because you think it looks cool. Alignment won't benefit your technical documents if it's used without purpose or an understanding of your audience.

Figure 3.11. Ineffective vs. Effective Alignment. Compare these examples of ineffective (A) and effective (B) alignment.

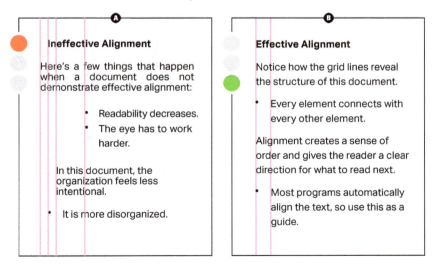

Grids

Grids are underlying structures used to help make sense of and create order within documents (figure 3.12). A grid also refers to how you set up the "rules" of your page to create strength and consistency in your design. Most documents follow a standard horizontal grid as in a memo or email. A vertical grid is often used in brochures or manuals, like the one Kenji is designing.

When you don't use a grid, you can end up with a variety of alignment planes. Grids are like guardrails that keep you on the right path. Take a look at these two résumé drafts. In the model of ineffective alignment in figure 3.13a, the text sends an unclear message to the user about where to look. In the model of effective alignment in figure 3.13b, the text guides the user's eyes down the page. The path that you create through your document is an important part of readability. By choosing the lines for your design, you either give your user a clear direction or create chaos.

Figure 3.12. Grids in Visual Design. Notice how the elements in Kenji's brochure line up with the grid, not just on a single page but throughout the entire document.

Figure 3.13. Ineffective vs. Effective Alignment. Scan these two versions of the same résumé. In example A, the text does not consistently align with the left margin. In example B, the alignment is consistent and predictable, making it easier to scan.

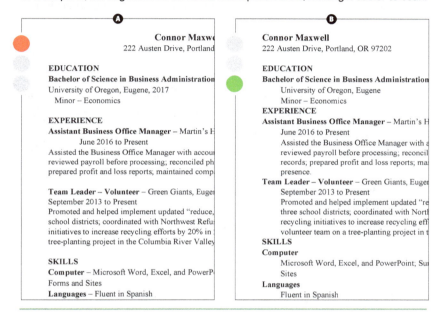

Design for Emphasis

Design in technical documents is functional, not decorative. Emphasis is one design principle that technical communicators employ to guide users. Specific design elements, such as white space, contrast, proximity, and headings, emphasize parts of a document for better understanding.

Consistency, consistency, consistency. You need to have clear, repeated design choices to give stability and structure to your document. The human brain works by pattern recognition, so much so that we sometimes see patterns where there aren't any. It's why a bunch of random craters on the moon look like a face to us. It's also why it's so hard to proofread your own work—you see what you expect to be there rather than what is.

By using repetition to create emphasis, you give the user confidence in interpreting a document. Repetition works by using any single element multiple times. For example, can you see how this page, this chapter, this textbook use repetition? Repetition is one way to create consistency and stability. Repetition can include color, shape, font, placement, line thickness, headers and footers, images, etc. Any element could be repeated to create this stabilizing effect. In Kenji's brochure, he uses the repetition of color, shape, and font to unify the document's design.

White Space

White space is any empty space surrounding figures, tables, visuals, or text. When used appropriately, it can direct a user's eye by isolating or emphasizing elements of document design (figure 3.14).

Don't neglect the invisible element of white space. You have probably centered the title of an academic paper, thus creating white space on each side. The white space draws attention to the title, as shown in figure 3.15.

Kenji uses white space in his brochure design around the date and location of the conference. The white space draws a user's eye to the visual element more quickly. White space guides a user through a document without distracting them with unnecessary information. Don't tell a user where to look—show them.

Contrast

In any document, you will see elements of marked space (text, images, shading, boxes, etc.) and unmarked space (white space). **Contrast** is an element of visual emphasis that makes a distinction between marked and unmarked space through intentional differences in size, color, or appearance.

If a document has too little contrast, its design will look muddy or indistinct. For instance, using multiple serif fonts in a single document looks odd because the fonts are too similar. Lack of variety leads to a lack of contrast, which reduces the document's effectiveness.

The same holds true if you use a 13-point font for your headings and 12-point font for your body text. Too much similarity in the heading styles means your user will not be able to tell them apart.

On the other hand, use contrasting elements selectively. You could also go too far and have too much contrast. You have likely seen some PowerPoint presentations with an overload of colors, fonts, text boxes, words, and animated graphics. Too much contrast ends up being more distracting than useful.

Figure 3.14. White Space for Readability. Use white space in technical documents to draw the user's eye to important information. These examples use placeholder text to emphasize the document's white space.

Figure 3.15. White Space for Emphasis. White space creates emphasis, but you may not notice it until it's missing (example A). In example B, white space can be found between the title and the text, between lines of text, between paragraphs, and even between words.

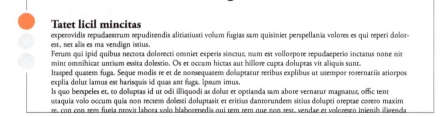

Notice the different kinds of white space in this document.

Proximity

Proximity has to do with where items are placed in relation to one another. Similar ideas should be close to each other. Dissimilar ideas should be farther apart. White space and grids separate and group different elements. When proximity is logical in a document, the physical closeness of similar types of information contributes to the user's understanding.

Kenji uses proximity in the brochure by keeping topics logically grouped. For example, the brochure will list contact information such as a phone number and a web address for the conference. Kenji will keep this information together in close proximity. Users expect all contact information to be in one place, not spread around the brochure. Kenji will also put similar content nearby, such as the address for the conference or directions. If Kenji hasn't created this kind of document before, he'll probably want to look at examples to get a sense of what's typical. That way, he'll use proximity in a way that's familiar to users.

Headings

Headings are words or phrases used as labels to group information within a document. The human brain likes to chunk information for understanding and retention, and headings allow the user to see where one section ends and a new one begins.

Headings can help you avoid the "wall of text" effect, as shown in figure 3.16a. The audience can use headings like road signs to guide them through a document. Headings make the information more organized and scannable. That way, users can glance through the sections and see the structure and relationship of information.

Headings also create a hierarchy of ideas, with main concepts receiving a top-level heading and subtopics receiving a lower level heading, or **subheading**. When headings and subheadings are done well, you should be able to read just those elements and understand the scope and sequence of the document. Figure 3.16b breaks up the "wall of text" using a first-level heading for "Recommendations." The document identifies three recommendations with a second-level heading.

Figure 3.16. "Wall of Text" vs. White Space. A document with no paragraph or section breaks like example A is hard to read. Example B shows the same text with headings and subheadings to break up the "wall of text" and increase readability.

A

Here are the recommendations. Given the relatively small size of the company, and its financial limitations in comparison to industry leaders, it would need to tread carefully into this area. The autonomous projects that require a driver's presence have been excluded from the discussion for safety and cost reasons. The following options are the most feasible at this time. The first option is partnership and pilot. "Corporation X" seems to be the U.S. automotive company most eager to release a fleet of AVs. They've developed partnerships with a few other companies in order to pilot their vehicles and are interested in using Tomorrow's Taxi Company as well. For the first year, they would provide the thirty-eight vehicles. Tomorrow's Taxi Company would be responsible for maintenance and energy costs, insurance, as well as monitoring staff. "Corporation X" would oversee any issues with the vehicles, make changes and upgrades, provide parts, and train staff. They would receive .5% of fare profits for this period for the use of the vehicles. At the end of the contract period, it could be renegotiated for another year if the technology is still needing improvement (which would likely be the case) or Tomorrow's Taxi Company would have the option to negotiate a purchase for the vehicles at a discounted rate. The second options is test vehicles. Under this scenario, Tomorrow's Taxi Company would lease-to-own two driverless vehicles from "Corporation A" at a rate of $20,000 per month for a period of six months. For the first six months, the corporation will handle the monitoring of the vehicles remotely. After the trial period, the company would have the option of extending for another six months or paying the balance on the vehicles and purchasing the monitoring equipment. If leased for an additional six months, an end-of-term buyout or return of the vehicles would be required. The third option is partial replacement. In this option, Tomorrow's Taxi Company would replace five of its vehicles that have the highest mileage with AVs and purchase the monitoring equipment at an estimated cost of $1,200,000. This would include replacing the two mini-vans currently used in the hilltop community, as these are experiencing significant

B

Recommendations

Here are the recommendations. Given the relatively small size of the company, and its financial limitations in comparison to industry leaders, it would need to tread carefully into this area. The autonomous projects that require a driver's presence have been excluded from the discussion for safety and cost reasons. The following options are the most feasible at this time:

Option One – Partnership and Pilot

"Corporation X" seems to be the U.S. automotive company most eager to release a fleet of AVs. They've developed partnerships with a few other companies in order to pilot their vehicles and are interested in using Tomorrow's Taxi Company as well. For the first year, they would provide the thirty-eight vehicles. Tomorrow's Taxi Company would be responsible for maintenance and energy costs, insurance, as well as monitoring staff. "Corporation X" would oversee any issues with the vehicles, make changes and upgrades, provide parts, and train staff. They would receive .5% of fare profits for this period for the use of the vehicles. At the end of the contract period, it could be renegotiated for another year if the technology is still needing improvement (which would likely be the case) or Tomorrow's Taxi Company would have the option to negotiate a purchase for the vehicles at a discounted rate.

Option Two – Test Vehicles

Under this scenario, Tomorrow's Taxi Company would lease-to-own two driverless vehicles from "Corporation A" at a rate of $20,000 per month for a period of six months. For the first six months, the corporation will handle the monitoring of the vehicles remotely. After the trial period, the company would have the option of extending for another six months or paying the balance on the vehicles and purchasing the monitoring equipment. If leased for an additional six months, an end-of-term buyout or return of the vehicles would be required.

Option Three – Partial Replacement

In this option, Tomorrow's Taxi Company would replace five of its vehicles that have the highest mileage with AVs and purchase the monitoring equipment at an estimated cost of $1,200,000. This would include replacing the two mini-vans currently used in the hilltop community, as these are experiencing significant

The contrasting size and font in the headings give users an immediate sense of the document's organization. Headings should be parallel to create consistency. **Parallelism** is the consistent balance and structure of words or phrases. For example, if you have a level one heading with two words, then your other level one headings should be of similar length. Users expect this level of consistency. Plus, if you want to create a table of contents for your document later, using consistent and parallel headings is essential. The design elements of white space, contrast, proximity, and headings all contribute to the organization of a document.

Design for Organization

See Chapter **6** for more organizational patterns.

The organizational structure for a document is determined by the audience. Take the brochure Kenji is working on as an example. The brochure covers the basic information the audience want to know: when the conference takes place, where it is happening, and who should attend. The brochure also includes information about New Delhi to encourage international travelers to visit.

Once you know your audience, you can create an effective technical document that uses logical patterns of organization to enhance readability. Start by using these preliminary questions as a guide:

» What is the purpose of this document?
» Who is my audience? What do they understand about the topic?
» What belongs where?
» How do I want the user to approach the material?
» What comes first? Next? Last?

Organization tends to follow a few familiar approaches. Some of the most common approaches include the use of headings, chronology, sequence, and a problem-solution format to organize information.

Chronological Organization

Chronological organization arranges information according to a progression of time. Explanations of how something happened, how something currently happens, or how something may happen in the future follow a precise time sequence to guide users.

For example, Kenji's previous market research showed that professional engineers are increasingly concerned over the safety of industrial control system cybersecurity. This type of cybersecurity is specialized to determine vulnerabilities not only for data, but also for the automated systems that engineers design and oversee remotely and on location. Without cybersecurity to outpace hackers or vandals, systems could go down, placing whole communities in danger. Kenji knows that Tavent is creating a presentation on cybersecurity for the upcoming conference. In the conference description on the company's web site, Kenji suggests Tavent guide users through the past, present, and future of engineering conferences (figure 3.17).

Figure 3.17. Chronological Organization Model. Compare the draft in example A with the text organized by chronology in example B. The time-based and specific language highlighted in the second example makes it easier for readers to understand the progression of events.

A

Draft with Limited Guidance

Cybersecurity used to be part of conference seminars on other issues. Cybersecurity discussion at conferences increased after threats. Automated solutions began to be included, but engineers have to maintain those systems. Engineers should be connected to cybersecurity consulting firms who provide maintenance services instead.

B

Draft Organized by Chronology

In past conferences, like those in the 1990s, cybersecurity was an emerging topic mentioned as part of other relatable issues. Increased threats in the 2000s created an urgency for cybersecurity discussion. Present industry conferences offer seminars on automated solutions that engineers are responsible for maintaining. A future conference could connect engineers to cybersecurity consulting firms who could provide best practices for protection instead.

Sequential Organization

Sequential organization is similar to a chronological pattern but arranges information according to a logical step-by-step sequence that describes a particular process (figure 3.18). Using this pattern of organization requires each section of information to represent a main step in an actual process.

Often these steps are written from the user's perspective (in this case the implied second-person pronoun "you"). This is done so that users can see themselves completing the process successfully. A user manual or set of instructions are the most common examples of this kind of organization.

Figure 3.18. Sequential Organization Model. Notice the use of parallelism in this example with the repetition of visual and verbal elements.

1 **Practice and prepare for the video interview**
→ Practice your answers to the anticipated questions.
→ Prepare notes to prompt you at a glance during the video interview.

2 **Dress yourself and your setting for business**
→ Dress nicely like you would for any interview and not just from the waist up.
→ Clear your background so viewers only see what you want them to see.

3 **Test equipment well in advance**
→ Make sure your webcam, audio, and connection are all working beforehand.
→ Connect to your interview session ahead of time to avoid delays from logging in at the last minute.

4 **Captivate your viewer**
→ Introduce yourself, continue to smile, and make eye contact with the webcam during the interview.
→ Answer questions to highlight your skills and speak clearly and with confidence.

5 **Follow up in style**
→ Send the interviewer a personal thank-you right away.
→ Check back with the interviewer in a few weeks.

Problem-Solution Organization

Effective **problem-solution organization** divides information into two main sections: one that describes a problem, and one that describes a solution. This kind of organization has standard sections. For instance, a document that uses problem-solution organization, such as a short or long report, will include a statement of issue, research, analysis, and recommendations for improvement. You may want to use problem-solution organization when your goal is to convince the user to support a certain course of action.

Types of Visuals

Elements of visual design help users understand and remember pertinent information in a document. Visual design often follows social conventions that make material more user-friendly. Common visuals include images, photographs, tables, charts, and graphs. Visuals balance out text, provide another format for understanding, and connect the visual learning areas of the brain.

See Chapter 2 for more on distorted or misleading information.

Tables, charts, and graphs are ways of displaying complex information in more accessible ways. You can think of each visual display as a tool with a specific use. Each tool gives us a unique way of understanding the information. If you choose the wrong visual display, you risk confusing the user or even misrepresenting the facts.

As a professional, your job is to make ethical choices in the visual presentation of data. Not doing so could cause a legal problem for you and your employer. Remember that visual data can be deliberately distorted, but it can also be misrepresented through sloppy work. In either case, the technical communicator is at fault, so pay attention to the accuracy and appropriateness of your visual communication.

Tables

Tables organize data visually in columns and rows for easy comparison. Tables can be used to organize numbers or words (figures 3.19a and 3.19b). When is it best to use a table? If your purpose is to show multiple facts and figures, a table might be your best bet. It's an efficient tool for organizing information so that the user can quickly scan and compare.

Figure 3.19a. Numerical Table. Kenji uses a numerical table to show conference attendance in recent years and its relationship to the registration fee. The table shows trends and allows for easy comparison of information.

Date of Conference	Attendance	Registration Fee
2014	5,216	$50
2015	6,844	$50
2016	4,988	$75
2017	7,176	$55
2018	8,000 (estimated)	$65

Figure 3.19b. Prose Table. Kenji provides a prose table that allows conference volunteers to troubleshoot the most common problems they might encounter during the event.

Problem	Contact	Solution
Conference room is locked	Conference center security	Post note on door until security arrives
Projector is not functioning/available	IT support	Inform presenter and wait until IT support arrives
Not enough chairs	Maintenance	Obtain additional chairs from hallway and assist with setup when maintenance arrives

Charts

Charts organize data to show relationships by using shapes, arrows, lines, and other design elements. There are many types of charts. Common examples include flowcharts, pie charts, and organizational charts (figure 3.20).

Pie charts are useful for displaying data for about six categories or fewer. Kenji chooses a pie chart to show the types of activities that will be available at the conference. Since Tavent is trying to increase its attendance numbers, Kenji breaks down the events as a way to persuade more people to attend.

Figure 3.20. Different Types of Charts. Different charts serve different needs. Flowcharts are a good choice to illustrate a sequence of events or process. Pie charts make it clear how all the parts add up to a whole. Organizational charts help us see who reports to whom and the hierarchy of roles and responsibilities within an institution or company.

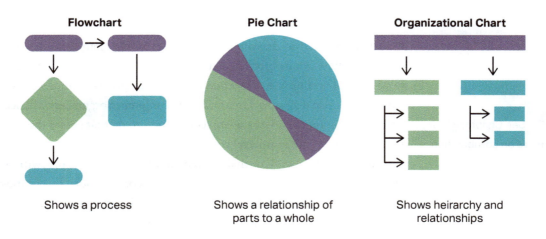

Flowchart	Pie Chart	Organizational Chart
Shows a process	Shows a relationship of parts to a whole	Shows heirarchy and relationships

Graphs

A **graph** translates numbers into shapes, shades, and patterns. Graphs can display approximate values, a main point about those values, or the relationship between the two. There are generally two types of graphs: bar graphs and line graphs.

In a simple **bar graph**, horizontal bars or vertical bars represent the number of items in a particular category. A bar graph shows trends or changes over time. In Kenji's example of a bar graph, he defines the category he wants to measure: the types of engineers who attended last year's conference (figure 3.21).

A bar graph that displays its information horizontally is sometimes called a **line graph**. Line graphs display how one part, or variable, is affected as another rises or falls. In Kenji's market research, his line graph shows how conference attendance fluctuated over the past four years (figure 3.22).

Illustrations and Diagrams

Illustrations rely on drawings and sketches rather than data or words. Technical illustrations often depict what words cannot. **Diagrams** show how parts of an object fit together. Common diagrams include cutaway, cluster, and cycle diagrams. Eventually, the conference organizers will need an illustration for their website and conference guide that shows the location of events in the multilevel convention center.

See Chapter 7 for more on collaboration.

Kenji didn't make all of this happen on his own. First, he met with his client, Tavent, to figure out their needs. In that conversation, Tavent most likely relayed the needs of their clients, the potential attendees. Once Kenji had information to work with, he then returned to his team to conduct market research. Next, he got his graphic designers on board to design the brochure and PDF. Lastly, he organized writers to create the content. As you can see, it takes a team of people to create reliable and trustworthy content.

Figure 3.21. Bar Graph. This simple bar graph shows who attended the conference. To show attendance over time, a multiple bar graph would have a bar for each year in each category.

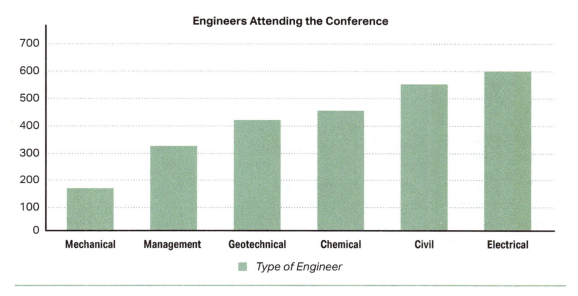

Figure 3.22. Line Graph. Kenji's line graph examines one data point: how many people attended the conference year by year. In a multiline graph, Kenji could examine the numbers of different types of engineers who attend year by year, assuming he has that data.

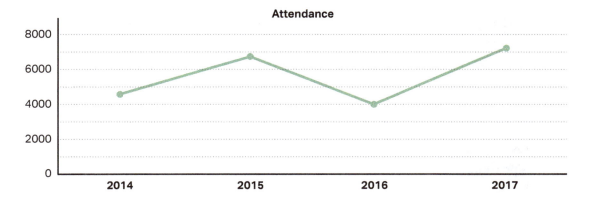

Case Study

Develop a Document Plan

This case study is an opportunity for you to put into practice what you've learned. Part of this chapter focuses on creating readable, useful documents so the user can find what they need and understand the message. Look at the following case study to consider how creating a document plan keep you on track and provides a consistent layout for the user:

> See Chapter **7** for more on workplace documents.

A document plan comes after you identify an issue in need of a solution. This process leads you to the purpose, audience, and message of your document. Think of a document plan like an outline that helps you organize your ideas.

Document plans come in different shapes and sizes. They can be as short as a basic memo or as long as a multipage report. In figure 3.23, Kenji begins with a statement of the issue (his purpose). After establishing his audience and purpose, Kenji asks himself the following questions:

» **Setting:** Where will the user most likely access the document? On the go? In their office?
» **Potential obstacles:** What are some potential problems of this document plan?
» **Format:** What kinds of text, visuals, and layout are most useful for conference goers?
» **Timing:** What design deadlines will keep the team on track?
» **Budget:** How might the available funds from the client affect the final product?

When planning is complete, Kenji is ready to write, test, revise, and proofread a document. Because he works with a team of people, this planning document prepares his team to be successful as they move into the next phase of the project.

Discussion

» Print and online documents have different requirements. What three features of design are vital for effective online design?
» Look up effective sentence and paragraph length for the type of document Kenji's team is preparing. What did you find?
» What message should Kenji and his team aim to convey to the audience?

Case Study, continued

Figure 3.23. Document Plan Model. A basic document plan can be as short as a memo.

Memorandum

To: Step-by-Step Planning Team   From: Kenji Y.
Date: 05/23/XX   Re: Design Plan for Tavent

Step-by-Step recently completed an audience analysis of Tavent's conference attendees in preparation for its online and print documents promoting their new event in India 20XX. This memorandum summarizes Tavent's findings and presents a plan of action. The full audience analysis can be found in the research marketing department's shared file.

Phase One: Audience, Purpose, Message
Primary Audience Summary: Conference attendees come from all areas of the globe. Most notably, there is a concentration of engineers in Russia, United States (U.S.), Iran, and Japan. Our findings show that Tavent's past conference attendees came mostly from the U.S. and Japan.

Purpose: The primary purpose for Tavent's online and print document is to promote a new engineering conference in India 20XX. The secondary purpose is to attract more engineers. Our challenge is creating a document that draws more engineers from Russia, Iran, and smaller markets. Tavent hopes to create a diverse, innovative conference in India.

Message: The key messages for the conference collateral include equity, education, and innovation.

Phase Two: Design Plans
Features: Both the print and online brochure will use Tavent company colors to maintain consistency in branding. In addition to promoting the conference proceedings, the collateral also needs to include travel and economic information for New Delhi.

Outline: Trifold print brochure. We will design the print version first and then make adjustments to it for online distribution.

Production Schedule: Print brochure must be available at the 20XX conference, one year in advance of the New Delhi conference. Early bird pricing should be included on both the print and online collateral by June 20XX.

Begin with a brief introduction.

Use headings to guide the reader.

The message depends on knowing the audience and purpose for the document.

Budgets and timelines may also be part of document plans.

Checklist for Document Design

Readability

- ☐ Have you chosen a readable font (most likely sans serif font for technical documents)?
- ☐ Do you vary your sentence length and avoid fragment sentences or sentences that are too lengthy?
- ☐ Have you used a grid to align elements within your document?

Organization

- ☐ Have you arranged your document according to user needs (such as chronologically or sequentially)?
- ☐ Is the organizational pattern easily detectable?
- ☐ Have you grouped content logically to help your user navigate?

Emphasis

- ☐ Have you used white, or empty, space to draw attention to certain design features?
- ☐ Have you used contrast selectively to avoid overwhelming the user?

Conclusion

To be successful in both page layout and design, you must skillfully blend the multiple components of document design, visual design, and organizational design. In all three areas, the key is to focus on the common goal of creating a document that meets the needs of users.

Remember that all design—whether it's technical communication or something as ordinary as a paper clip—seeks to simplify the user's experience. The principles of readability, emphasis, organization, and visual supplementation make it easy to design documents that users understand.

In a technical communication course, you will likely be graded on your individual understanding of the concepts in this textbook as well as the ideas presented by your professor. Beyond the classroom, however, technical communication is a complex field that requires an understanding of page layout and content as well as thoughtful, purposeful design features. Just as words and images must work together in a technical document to create meaning, you need to sharpen your ability to work with tools and in teams to create the best possible product.

Chapter 4

Multimodal and Multimedia Communication

Abstract: Multimodal documents use a combination of written, auditory, spatial, gestural, and visual choices to help convey and deepen a document's message. Technical communication requires that you understand the choices a writer/designer makes and be able to explain the chosen output. In professional workplaces, your boss will expect you to be able to explain why and how you chose a certain mode of expression for the information and how it benefits the end user. To be successful, you will need to understand and apply a range of communication modes and media.

Looking Ahead

1. Why Multimodal and Multimedia Communication Matter
2. Multimodal Communication at Work
3. Rhetorical Awareness and Digital Literacy
4. Output Options

Key Terms

» digital literacy
» medium, media
» mode
» multimedia communication
» multimodal communication
» rhetoric
» rhetorical awareness

Why Multimodal and Multimedia Communication Matter

Think about the many ways you absorb information daily. Maybe you read your news online, stay updated on family and friends via social media, or get tips on how to perfect your standup routine through podcasts. You interact with different modes—simply put, a variety of communication methods—when you take in ideas through multiple channels, such as text, images, videos, or podcasts.

Multimodal communication is a system of relaying information that includes modes that are gestural (movements), linguistic (words), aural (sounds), visual (images), and spatial (physical arrangement) (figure 4.1). Because the use of multiple modes of communication is so common, you may not recognize it at first. Ever play charades? Or create a home budget on a spreadsheet to track your monthly expenses? Congrats. You're already living a multimodal life.

The digital age has produced an increased interest in multimodal communication. Your phone, computer, and social media accounts provide

Figure 4.1. Types of Multimodal Communication. Multimodal communication mixes together different ways of conveying meaning, mainly linguistic, aural, visual, spatial, and gestural modes.

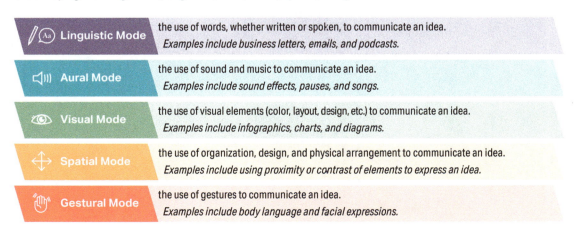

Linguistic Mode — the use of words, whether written or spoken, to communicate an idea. *Examples include business letters, emails, and podcasts.*

Aural Mode — the use of sound and music to communicate an idea. *Examples include sound effects, pauses, and songs.*

Visual Mode — the use of visual elements (color, layout, design, etc.) to communicate an idea. *Examples include infographics, charts, and diagrams.*

Spatial Mode — the use of organization, design, and physical arrangement to communicate an idea. *Examples include using proximity or contrast of elements to express an idea.*

Gestural Mode — the use of gestures to communicate an idea. *Examples include body language and facial expressions.*

different avenues for you to make meaning. Even if you don't own a cell phone or personal computer, you still participate in a multimodal life because it's all around you.

Multimodal and Multimedia Communication Defined

Multimodal and multimedia sound similar, but don't let that fool you. **Multimodal communication** refers to how meaning is created and experienced through a combination of modes: visual, aural, linguistic, spatial, and gestural. **Multimedia communication** refers to the final product that serves as a container (or medium) for the communication. The difference between mode and medium is subtle, and both terms can be applied equally to a single form of communication. The next two sections will help break down the difference.

Modes

Mode is the style or way content is presented. Mode can be text, images, sound, movement, or anything that creates and conveys meaning. Interpretive dance is an example of a movement-based mode that is useful for describing complex emotional landscapes difficult to put into words. However, this mode would be useless when trying to efficiently describe the safety precautions necessary to operate a forklift.

Multimodal describes using more than one mode at the same time. For example, if you give a presentation to your class, you'll likely write an outline beforehand (text). You'll vocally deliver your ideas in front of the class (speech). To make your presentation interesting, you add in a visual aid or animated slides with sound effects (color, image, movement, sound). You use the room as your stage and invite audience participation (gesture, movement). You might not have thought about it directly in these terms, but by incorporating all these elements, you've given a multimodal presentation.

Multimodal communication is everywhere. For example, a visit to any modern retail space will bombard you with messages in multiple modes. When you pass a store playing loud, danceable music, that store is using the aural mode to indicate the store is full of fresh styles you'd feel good wearing. If you walk in, employees use the gestural mode of smiling at you to indicate you are welcome and belong in the store. The spatial mode is used to arrange the store so you will pass points of interest in a specific order—the clearance rack is never by the door, and accessories are on the way to the register. Meanwhile, you experience the visual mode through images of people looking good in the latest fashions. Even the flattering mirror shape and lighting in the fitting rooms are designed for a purpose. You encounter the textual mode via exciting buzzwords, information, and price tags. Careful study goes into how the store communicates its brand. Take a moment to consider what multimodal elements might persuade you to choose one store over another.

The pervasiveness of multimodal communication means that modern audiences expect it. As a technical communicator, you should be ready to satisfy this expectation. As a student and developing professional, you should understand why the choices you make about the process (your mode) and how the finished product is distributed (your medium) both create meaning for the user. This sequence requires you to focus specifically on how to transfer meaning to a user. You must think about what your audience understands and how they need to receive new information to make it useful.

Medium/Media

A **medium** is a means of transmitting information or data. A résumé is an example of a medium. So is a website. Both examples deliver information. They are the result of a process. **Media** is the plural of medium. Multimedia refers to more than one medium being used in the same application. For example, a multimedia presentation might feature printed handouts, video clips from a documentary, or audio recordings from an album.

In this chapter we use the term multimodal to include both the process of creating meaning through multiple modes of communication and the final product or output (medium). Understand that when you read multimodal from now on, the idea of multimedia is implied. When you apply multimodal communication, it will inevitably take the form of a medium.

Every piece of communication has a mode and a medium. For example, the TED Talk by Suchitra Krishnan-Sarin called "What You Should Know about Vaping and E-Cigarettes" combines linguistic, aural, visual and gestural modes, but it also has a medium (online digital video).[1] You might find it helpful to think of mode as the larger, general category and medium as the narrower, specific category (figure 4.2).

In this TED Talk, Krishnan-Sarin talks about the feeling of raising children, her expertise studying the biology of addiction, and the popularity of vaping while debunking myths about the safety of e-cigarettes. Her speech combines linguistic and aural modes. Her slides use the visual mode by showing different types of vapes, diagrams of e-cigarettes, and the definition of "chasing smoke." Her body language, facial expressions, and hand movements are subtle, but these do represent the gestural mode—especially when Krishnan-Sarin makes an "opening the door" gesture with her hand while presenting a new slide. The video offers an excellent example of a multimodal approach to presenting a technical topic to a broad audience.

Figure 4.2. Modes and Media. The five modes can be combined in countless ways to create a variety of media types. These media types include brochures, presentations, websites, podcasts, and many other output options that communicate an idea to the user.

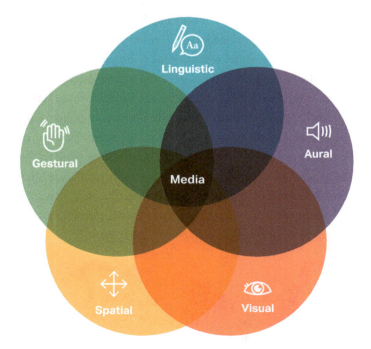

Multimodal Communication at Work

See Chapter 1 for more on the Problem-Solution Framework.

As always, the relationship between audience, purpose, and message should move the communicator from problem to solution. Let's examine a student's transition to the professional world, for example. Prathita recently graduated and accepted a forensic psychologist position at a community health organization that offers mental health access and services to the county's citizens.

In her new role, Prathita notices her manager doesn't provide guidelines like her professors did. Her job responsibilities include offering solutions that will provide people with the care they need during a crisis, rather than lead to unnecessary arrests. She is expected to present the data at an upcoming meeting. Where her professor would have offered Prathita detailed instructions and considerable help, Prathita's boss leaves her to work out solutions by herself. To be successful, Prathita needs to be resourceful and collaborate with her colleagues.

Prathita's position is funded by a grant, which means her department needs to show a return on investment (ROI) to remain funded in the future. She must report on the number of calls her department receives per month and the type of services provided. Based on her review of previous reports, Prathita understands that her report needs to be presented visually and in writing. In other words, her reports need to be multimodal.

Prathita must also consider her audience: a diverse set of stakeholders, like the health organization that hired her, the county police department, and the funders. The purpose of the report is to convince her manager, funders, and other stakeholders that her work and the work of the health organization remains valuable and allows the county to allocate funds where they are most needed. As a result, her report must be data-driven. Because this particular report must also take a multimodal format, Prathita needs to find a variety of methods to engage her audience.

Rhetorical Awareness and Digital Literacy

Technical communicators make decisions about how they will communicate rhetorically. **Rhetoric** is the study of how to best communicate, especially when justifying or validating a concept is involved.

Think about how you get someone to do something. If you want a new stop sign installed on your street, how will you make that happen? Rhetoric is the communication choices you make to convince the city that a stop sign is needed. You've probably come across the rhetorical terms of ethos, logos, and pathos in other writing classes. As a refresher, these terms represent three ways to convince an audience by appealing to ethics (ethos), logic (logos), or emotion (pathos). To use these rhetorical appeals effectively, you must understand your purpose and audience.

Rhetorical awareness means that you think about how to make choices based on the purpose of the document and its audience. Ask yourself what combination of modes will create the most meaning? As a professional, your responsibilities will include production and distribution choices. In the workplace, you most likely won't be asked to explain rhetorical principles. Instead, you may be asked to explain the medium (or communication platform) you chose. Your employer isn't as interested in why you chose certain colors for your civil engineering graphic, but they are interested in the usefulness and cost of an interactive blog you created for a neighborhood roadway restructuring project. Rhetorical awareness means understanding the situation and the needs that arise.

Digital Literacy

Digital literacy is your ability to locate, interpret, or generate information using a range of digital and online tools. Communication in the twenty-first century goes far beyond paper documents. Cell phones, tablets, computers, games, and virtual reality have changed the way people communicate and make meaning. If you want to remain relevant, you need to understand how to use and create multimodal content.

Digital literacy includes six essential skills:

» Knowing how to operate a computer or other digital device
» Knowing how to use a range of hardware and software
» Knowing how to keep your information secure online
» Knowing how to conduct an online search
» Knowing how to evaluate online information
» Knowing how to communicate and collaborate online

Hiller A. Spires, a professor of literacy and technology at North Carolina State University, provides another way to look at digital literacy. Spires says you can think of digital literacy as having three steps: 1) finding and consuming digital content, 2) creating digital content, and 3) communicating or sharing that content.[2] Each step of digital literacy contributes to the next (figure 4.3).

Figure 4.3. Digital Literacy Skills. Digital literacy requires three distinct skill sets.

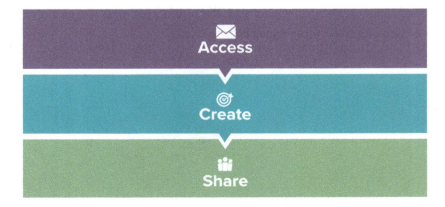

In preparing for her new job, Prathita thinks about the essential skills of digital literacy and concludes that she lacks expertise in keeping information secure online. As a result, she does some research to learn more about how to improve the security of her information, as well as the sensitive information belonging to her department. She finds ways to improve her passwords, to add login securities, and to save sensitive data more securely.

Because technology moves so quickly, it's inevitable that you will discover gaps in your technical skills even if you grew up knowing how to code computers or hack into your sister's Twitter account. Digital literacy is not just about knowing how to use these tools. It's also knowing when and how to use them, how to evaluate the information you receive, and how to translate this information into forms that are useful for others. By keeping up with changes in technology, you can build your capacity for multimodal communication.

Subscribing to publications that discuss new technologies is also a good way to keep yourself from falling behind in your field. Continuing education courses give you a safe way to experiment with new technology while guided by experts. As misinformation becomes more prevalent and convincing, remembering to triple-verify facts via the most credible, relevant sources should always be your top priority as a technical communicator.

Output Options

Many types of output (communication platforms) exist, and the options continue to grow and change at a rapid rate. This section discusses why and how you might choose one or more to create meaning for the user. This section offers a sampling of what might be at your disposal for each mode.

See Chapter 12 for more on staying current in your field.

Choose your medium thoughtfully. As a technical communicator, you need to make users confident in their interpretation of the information. You can build this confidence by using media in ways that are consistent with a user's experience. The more you look at good examples of media application and familiarize yourself with what makes them effective, the greater the likelihood that users will understand your documents. The output options covered in this section are photography, podcasts, social media, presentations, virtual reality, and websites.

Photography

Medium: photography
Mode: visual, spatial

At first glance, photography might appear to be only a mode of visual communication. With further exploration, however, you can see that photography also references the spatial mode by using physical arrangement to communicate an idea. Photography also provides an example of how media can be interpreted by different people. You may look at a photograph of a sunset and think it's romantic. Your neighbor may look at the same photograph and see darkness encroaching.

While photographs can add interest and emphasis to a document, they can also provide specific technical information (figure 4.4). In a recipe, a photograph can tell the user how the lasagna should look when it's finished. In a user's manual, a photograph of the toaster oven can be labeled to show its parts and functions. In assembly instructions, photographs can provide step-by-step guidance for the assembly of your new bookcase. In technical definitions or indexes, a photograph can aid in visual identification of a specific item.

Figure 4.4. Image as Technical Communication. This photograph communicates technical information that the user needs to complete a specific action.

Podcasts

Podcasts are collections of digital audio files that can be accessed online or downloaded to a smartphone, tablet, or computer. They represent the aural mode.

Podcasts come in many different genres (categories), including fiction, nonfiction, news, humor, how-to, and more. Podcasts are generally recorded and edited, but they can also be performed and recorded live for an audience. There are even podcasts about technical communication, such as "I'd Rather Be Writing," a podcast by Tom Johnson that explores trends and innovations in technical communication (figure 4.5).

Though a podcast is mainly an auditory experience for the listener, its creation is multimodal. Most podcasts have formats and recurring segments that listeners come to expect. These components are frequently outlined and planned in advance. Online show notes are paired with the podcast to document the show's content and to allow for additional information, links, images, and interactivity.

Medium: podcasts
Mode: aural

Figure 4.5. Screenshot of Show Notes. This is an example of show notes from "I'd Rather Be Writing." While the podcast is an example of the aural mode, the show notes use both the visual and linguistic modes to summarize the show, highlight key content within the show, and provide a link to the audio file. Used with permission.

Social Media

Social media includes websites and applications that allow users to generate and share content in highly interactive online environments. The most common social media site is Facebook with 1.79 billion daily active users, according to their publicly available 2020 second-quarter earnings statement.[3]

When you use social media professionally, review your company's social media policy and guidelines. Certain social media sites are better suited for certain kinds of content. As always, select the platform that best suits your purpose, audience, and message (figure 4.6).

Prathita, like most people, is familiar with posting content to Facebook and Instagram about her personal and professional experiences. She receives an amusing call from a client while at work and pulls out her phone to record the story. *Wait a minute*, she thinks, *This is a new job. I should check the policy first.* After looking at the social media policy, she realizes that talking about clients online violates her confidentiality agreement and she might get in trouble for being on social media while on the clock. Her boss, Annabelle, is already one of her Facebook friends and a stickler about not posting anything before 5 p.m.

Figure 4.6. Communicating through Social Media. Social media platforms may change, but the big three elements of technical communication (audience, purpose, and message) remain essential for anyone who wants to communicate effectively. This chart gives an example of why you might choose one platform over another based on these three important elements.

Platform	Audience	Purpose	Message
Instagram	Potential customer	Image sharing	New product announcement
YouTube	New customer	Video sharing	Step-by-step assembly demonstration
Twitter	Existing customer	Content sharing	Software update with link

Presentations

A presentation is a verbal performance that is delivered to an audience, either in person or virtually. A presentation might happen as part of a weekly production meeting. Or it may be an external meeting to keep stakeholders informed. You may give a presentation using a web conferencing tool to an audience you will never meet. Many presentation styles exist, but most incorporate a combination of modes, including spoken and written language, visuals, sounds, gestures, and movement.

See Chapter ⑦ for more on workplace communication.

One familiar and accessible form of presentation is the TED Talk. TED (short for "technology, education, and design") is a nonprofit organization known for its idea-based talks and conferences. One reason these talks are so popular is that they distill complex and timely topics into segments of fewer than twenty minutes. This brief format makes them easy to share. According to Chris Anderson, curator of TED, the common element that all successful talks share is a singular new idea, which he describes as "a pattern of information that helps you understand and navigate the world."[4]

Here are four tips from Anderson for giving a successful presentation:

- » Limit your talk to one major idea.
- » Give your listeners a reason to care.
- » Use familiar concepts.
- » Make your idea worth sharing.

You might never give a TED Talk, but the same principles apply to a presentation you might give to a room of four people or in a virtual meeting. Limit the presentation to one major idea. Otherwise, it's too easy to ramble about irrelevant issues.

All technical communication puts the user first, so it's your job to make sure the audience has a reason to care. If the content is complex and your audience is unfamiliar with the topic, look for ways to link your content to topics and experience your audience knows. Finally, you should be confident the idea is important enough, developed enough, and worthy enough to share.

Virtual Reality

Medium: virtual reality
Modes: visual,
aural, gestural,
spatial, and linguistic

Virtual reality is a computer-generated environment that creates the illusion that the user is somewhere else. Visual and auditory stimuli produced by the program via a headset combines with the user's physical and sensory experience to create an immersive and believable setting (figure 4.7).

Just as the invention of the telephone allowed for communication to take place between individuals in separate locations, virtual reality communicates the details of a distant (or even fictional) location to recreate it in the user's perceptions. In practice, virtual reality could be used to simulate just about any activity or model any scenario. If once you had to read the instructions to figure out how to use your new blender, now you can don a headset and see a master chef assemble it and use it in your own kitchen. Virtual reality is multimodal communication to the extreme, and its use for the purpose of technical communication is still being explored.

Figure 4.7. Virtual Reality as Technical Communication. Imagine being able to see complex procedures conducted in three dimensions. Virtual reality has potential within technical communication for training, usability testing, demonstrations, simulations, and walk-throughs.

Websites

The first webpage went live in 1991,[5] and since then websites have evolved in design, functionality, and mobile responsiveness. You can chat with someone on the other side of the world. You can order products and have them delivered in a matter of days. You can watch a video on your phone. The amount of data created and uploaded each day is staggering. Every second, 40,000 Google searches are conducted.[6] About three-quarters of the U.S. population has broadband internet access as of 2019.[7]

When you create content for an online medium, you must think about how people interact with information. Because most people scan websites rather than read them, websites are often organized visually and spatially (figure 4.8). Web-based text needs to be short, clear, and well-balanced with images.

Medium: websites
Modes: visual, aural, gestural, spatial, and linguistic

Figure 4.8. Website as Technical Communication. The homepage for Chemeketa Community College's website gives visitors everything they need to know at a glance.

The images used on the website's homepage communicate volumes about the college's values and mission without saying a word.

Chemeketa's logo occurs in the top left quadrant of the webpage and is used as a watermark in the center of the page to give added emphasis to the text.

The top three navigation links correspond to different users of the site. The orange tabs are designed for students and donors. The second level of navigation links are for specific audiences. The bottom navigation links appeal to general audiences or prospective students.

Case Study

From Student to Professional

This case study is an opportunity for you to put into practice what you've learned. Part of this chapter focuses on the difference between creating documents for a class or for the workplace. Look at the following case study to consider how the expectations of the user affect the rhetorical situation and mode of the end product:

Let's take a look at how one student navigates the world of multimodal communication as she transitions from graduate student to employee. Prathita is about to complete her education in forensic psychology. In her last quarter of graduate school, the instructor assigns a presentation about the job duties of forensic psychologists to evaluate students' software skills and their ability to create effective visual presentations.

Prathita will be graded on the layout, design, and visuals she chooses for her presentation. She focuses on creating a consistent layout with simple but effective design elements and relevant visuals. Her grade is based on her ability to meet the requirements of the assignment.

During this same quarter, Prathita was hired by a local community health organization offering mental health access and services to the county's citizens. Prathita's transition from student to working professional will focus on her ability to expand projects from development choices like layout and design of projects to include distribution choices. She will need to understand her audience's needs. For instance, Prathita's first work project is a multimedia presentation for a departmental meeting that takes place via a video conference for all counties in the state.

Discussion

» Describe the rhetorical situation of her school project and her work project. Where do the projects differ? Where are they the same?
» Discuss how her first work project will be different from her final project in graduate school.
» What modes would be most effective for her work presentation? Why?

Checklist for Digital Literacy

Finding Information

- [ ] Is the source trustworthy?
- [ ] Who created the content and why?
- [ ] What techniques are being used to attract attention to the content?

Evaluating Information

- [ ] Does the source allow some kind of engagement with the content, other than just reading?
- [ ] Who is the intended audience and how can you tell?
- [ ] Is there bias in the content? Are you accessing information with your own bias?

Composing Information

- [ ] Is your content communicated responsibly and safely for all users?
- [ ] Are there any potential consequences to the way people engage with content you create?
- [ ] Is the content equitable and accessible?

Conclusion

The ever-expanding digital landscape is rooted in the concept of multimodality. If you didn't grow up with this constant flow of information and changing technology, the pace of it all can feel daunting. Even digital natives—individuals who have used technology from an early age—sometimes have a hard time keeping up. Whenever you feel awash in the sea of information overload, use the fundamentals of technical communication as your life raft. Just apply these classic ideas to new contexts and situations and you're ready for a multimodal world.

Notes

1. Suchitra Krishnan-Sarin, "What You Should Know About Vaping and E-Cigarettes," TEDMED 2018, November 2018, video, 14:21, https://www.ted.com/talks/suchitra _krishnan_sarin_what_you_should_know_about_vaping_and_e_cigarettes.

2. Hiller A. Spires, "Digital Literacies and Learning: Designing a Path Forward," Friday Institute White Paper Series, 2012, https://www.fi.ncsu.edu/wp-content/uploads/2013/05/digital -literacies-and-learning.pdf.

3. "Facebook Reports Second Quarter 2020 Results," Facebook Investor Relations, July 30, 2020, https://investor.fb.com/investor-news/press-release-details/2020/Facebook-Reports-Second -Quarter-2020-Results/default.aspx.

4. Chris Anderson, "TED's Secret to Great Public Speaking," TED Studio, March 2016, video, 7:48, https://www.ted.com/talks/chris_anderson_ted_s_secret_to_great_public_speaking.

5. Alyson Shontell, "Flashback: This Is What the First-Ever Website Looked Like," *Business Insider* online, June 29, 2011, https://www.businessinsider.com/flashback-this-is-what-the-first-website -ever-looked-like-2011-6.

6. "Google Search Statistics," *Internet Live Stats*, accessed August 26, 2020, https://www.internetlivestats.com/google-search-statistics/.

7. "Internet/Broadband Fact Sheet," *Pew Research Center* online, Internet & Technology, June 12, 2019, https://www.pewresearch.org/internet/fact-sheet/internet-broadband/.

Chapter 5

Research Methods for Technical Communication

Abstract: The fundamentals of research for technical documents are similar to the research strategies for other types of documents. However, technical communication tends to focus less on research for rhetorical goals and more on research for analytical purposes. The primary purpose of research in technical communication is to test a hypothesis. Because research quite often turns up information that disproves your hypothesis, you should allow ample time to revise and narrow your research question. Even a failed hypothesis can produce useful results. Effective research for most technical documents includes analyzing data and studies. Efficient researchers keep detailed notes so they can cite their sources and avoid claims of plagiarism or theft of intellectual property. Ultimately, this chapter guides you in taking responsibility for finding answers, evaluating the answers proposed by others, and delivering the best answers to those who need them to make informed decisions.

Looking Ahead

1. Why Research Matters
2. Steps for Research
3. Primary and Secondary Research
4. Using Sources
5. Effectively
6. Citing Sources
7. Intellectual Property
8. Advanced Research

Key Terms

- » analysis
- » APA style
- » Chicago style
- » copyright law
- » fair use
- » hypothesis
- » IEEE style
- » intellectual property
- » MLA style
- » paraphrase
- » peer review
- » plagiarism
- » primary research
- » public domain
- » secondary research
- » subject matter expert (SME)
- » summary
- » work for hire

Why Research Matters

As a college student, you've likely had considerable experience with conducting research. This prior experience forms a good basis for your research skills as a technical communicator. This chapter reinforces what you already know about effective research, but it takes research a step farther by showing you what's important and unique in technical communication.

Here's a scenario to demonstrate how one might conduct the research process in a professional setting. Jessamyn works for Tomorrow's Taxi Company. Jessamyn's job involves researching issues for the company and putting her findings into technical documents so that her boss can make informed decisions. Jessamyn's boss has asked her to research the feasibility of adding electric cars to the company's fleet of vehicles. Jessamyn needs to do considerable research about the cost associated with adding these vehicles before she can create her deliverable (a feasibility report).

See Chapter **11** for more on feasibility reports.

Research Defined

Fundamentally, research functions the same way regardless of your field or industry: you look for the most credible information to arrive at a conclusion, and you present your findings in a format that meets the audience's needs and expectations. The techniques you employ while researching an argumentative academic paper and those you use for technical communication are similar enough that you shouldn't feel lost while reading this chapter. There are, however, some differences to take into consideration.

Ideological concerns or biases do not have a place in research for technical documents. In other words, technical communication does not have room for opinions. In academic writing, your informed opinion often contributes to your conclusions. Frequently, the goals of academic writing are rhetorical—that means your purpose is to convince an audience to see the topic your way. Technical communicators rely less on rhetoric and more on research to lead decision-makers to the best conclusion. For technical communicators, sometimes a failed hypothesis is as useful as a successful one. The technical document you create may recommend further study or taking a different course of action than originally planned.

Research at Work

Back at Tomorrow's Taxi Company, Jessamyn begins her research on electric vehicles (EVs). Her company has identified an issue with its fleet of cars. Its gasoline-powered cars break down frequently and contribute to air pollution, costing the company money on repairs and missed revenue opportunities. Jessamyn's boss asks her to investigate whether adding EVs to the company's fleet is affordable and reasonable.

A professional approach to this assignment means that Jessamyn does not simply decide on the recommendation she thinks is best and find evidence to back up that opinion. Jessamyn needs to do considerable research to find out whether adding EVs makes sense. Rather than leaping to conclusions, she needs to allow the data to guide her toward her conclusions.

Like most people, Jessamyn has a personal preference when it comes to gas-powered versus electric-powered vehicles. These preferences came into play when she decided to purchase her own car. When it comes to the recommendations she offers Tomorrow's Taxi Company, however, she needs to be guided by data. She needs to consider what's better for the company. Jessamyn's job in this situation is to help her company make the best decision. Ultimately, she is solving a problem.

Research and Problem-Solution Writing

Jessamyn thinks through the case of Tomorrow's Taxi Company. The problem is the cost of vehicle maintenance and fuel. Quarterly expense reports confirm this problem. Jessamyn's boss has narrowed the possible solutions that she could consider by directing her to investigate EVs. So, her researchable problem has been made more particular—she must find a solution in the form of an answer. Her answer must justify whether or not EVs will save money.

Jessamyn needs to consider the interplay between audience, purpose, and message as she researches a solution. The members of her audience are decision-makers in her business, including her boss, other administrators, and investors in the company. These people aren't interested in Jessamyn's feelings on the subject. Instead, they want to be convinced by the evidence Jessamyn finds.

The purpose of Jessamyn's project is to provide administrators with a recommendation regarding adding EVs to the taxi fleet. This purpose focuses her research. Even if she finds other useful information along the way, she needs to avoid getting off topic in her research. She also needs to avoid biased or inaccurate information sources, even if they confirm her suspicions about whether EVs are a good idea.

Jessamyn's message needs to be a clear recommendation in favor of or against using EVs. She needs to present this message in a way that is rooted in research. Her recommendation will only be respected by her audience if she can demonstrate that it is warranted by data and clear reasoning.

Having established how to solve the problem, Jessamyn can proceed with her research. She follows a set of steps in her research that she learned as a college student. The next section explains how you can follow a similar methodology for how you conduct research.

Steps for Research

As in Jessamyn's scenario, you need to determine what problem needs solving. Sometimes this is easy: you receive an assignment from your boss or instructor. The assignment has a clear outcome, such as "determine the feasibility of switching to electric vehicles."

In your case, you may get an open-ended research project from your instructor. Your instructor may assign you a topic or give you the option to decide what you want to research and why. In your professional life, you can expect to encounter any combination of reasons to do research that range from bidding on a work contract to submitting a business plan to qualify for a small business loan. Doing effective, honest, ethical research is a part of everyday life in the professional world.

Step One: Observe the Problem

To begin the research process, you need to spend some time observing the problem. For the purposes of illustration, let's say your problem is that you're waking up exhausted. During this stage of research, you begin by making observations and collecting data about your energy levels. Are there certain days you're more tired than others? Does it correspond to the number of hours you slept or to the amount of caffeine you did or did not consume? Do you notice that you're more tired on the mornings you don't eat breakfast? Does the room's temperature affect your sleep quality? Are you more or less tired when you exercise the day before? What about screen time? Do you get better sleep when you turn off all screens an hour before bedtime? Do you even have a bedtime?

As you collect your data points, look for patterns, trends, correlations, or changes that can lead you toward the formulation of a more precise research question. Determining research questions can help guide your research.

Step Two: Form a Question

To refine your research, you need to come up with a specific question. You could ask yourself, "Why am I so exhausted?" but you're unlikely to find research that will be able to provide a suitable answer. Effective research requires an appropriate level of focus. You will need to narrow your question to one of the data points you observed earlier. Let's say that you noticed you woke up multiple times during the night because you were too hot. You can narrow your focus by pairing your original idea (sleep quality) with a more specific subtopic (room temperature). A researchable question might then be, "What is the ideal temperature for better-quality sleep?" The process of defining a narrow, researchable question will keep you from getting distracted by too many possibilities.

Step Three: Propose an Answer

Those of you familiar with the scientific method probably see where we're going with this. You need to form a **hypothesis**, an educated guess that will then be tested. At this stage, you come up with an answer to your research question based on what you may already know or think you know about the issue you're researching. Your experience of tossing and turning at night suggests that room temperature plays a part in sleep quality. Your hypothesis, then, is this: "Cooler room temperatures contribute to better sleep quality." The value of a hypothesis is that it tells you what you are and, more importantly, are *not* researching.

Step Four: Test the Hypothesis

This step is the biggest difference between academic research and research in technical communication. In technical fields, the purpose of research is to test the hypothesis. Your goal isn't to prove you are right. Instead, the goal is to determine whether the proposed answer is supported by evidence. To test your hypothesis about cooler room temperatures, you could turn down your thermostat and keep a record of your sleep quality over the course of several weeks. Better yet, you can find a researcher who has already tested this hypothesis on a larger group of test subjects and published the results in a peer-reviewed journal.

In your search for a better night's sleep, your preliminary research may lead you to change your hypothesis. Maybe sleep quality is equally impacted by the room's temperature and the amount of artificial light in the room. In this scenario, you may find that you get better results if your research explores two factors rather than one.

Step Five: Draw Conclusions

The next two steps of the research process require discipline. Many of us don't like being wrong, which can lead to questionable choices during the research process. As in all areas of communication, honesty is the best policy. If you manipulate or exaggerate your results, you may have to deal with serious consequences.

If your hypothesis doesn't work out, simply say so and suggest an alternative solution or new research angle. The user will then act based on your well-informed advice.

Preliminary Research

Dividing research into two stages can save time. The first stage is preliminary research, which you use to accomplish the following:

» **Find useful search terms.** Take time to learn how professionals in the field refer to what you're looking for. This is one of the better uses for sites like Wikipedia—you can scroll through and look for terms that might get better search results. For instance, the more common term for myocardial infarction is heart attack. If you're looking for recent medical research on heart attacks using the search term "heart attack," you will get limited results. You have to figure out how experts in the field talk about the topic. Create a list of key terms.

» **Take your research to the next level.** It's rare to come up with a research project that hasn't already been explored on some level. If you find yourself at a loss for sources, it probably has more to do with what's going into the search bar than a lack of existing research covering the topic. If this is the case, seek the help of a reference librarian. Most reference librarians staff 24-7 chat sessions, have direct lines, or respond to emails. You can find this feature by going to your library's homepage.

» **Collect possible sources.** Since most topics have already been researched by someone else, you can use this preliminary stage to collect the material you want to examine more closely. Use credible sources that list the references they used.

Chances are you will read more than what you directly reference in your report. Jessamyn will collect information about the problem of gasoline vehicles and potential solutions of EVs so she can attain a level of expertise on the topic. That means she will read all types of articles, trade journals, and scholarly sources to inform her recommendation.

Preliminary research involves a lot of skimming. You can find tips for skimming documents later in this chapter. This stage involves concentrating on secondary sources and determining whether you will need to do any field work.

Final Research

The final research stage is where you do a deep dive. Effective research is like increasing the power on a microscope to see the details of your project better. You'll need to have a solid grasp of the big picture alongside the granular details because you'll probably be tasked with explaining your findings when necessary. If your document is longer, like a technical research report, part of the document will show the granular details you discovered through your research.

As you develop your research, you will most likely encounter new questions, new answers, and further research. This is where having a focused question and clear hypothesis comes in handy: it organizes your research according to relevance.

Step Six: Narrow the Research Topic

Let's revisit the sleep-quality scenario and imagine that you are waking up tired every day. This may or may not be a technical issue, but it provides some background for a possible technical research topic.

First, you need to observe the pattern and eliminate multiple possibilities. Your bedroom temperature runs hot in the summer and cold in the winter. This variation doesn't affect your sleep quality. Beyond your first-hand experience, your preliminary research on room temperature and artificial light didn't produce useful results.

Your next step is to narrow your topic. The common factor in all your sleepless nights has been your old mattress—the one you've been sleeping on for most of your adult life. Now you have a somewhat technical research question: "What mattress will give me the best rest for my money?"

This topic triangle can help you see how to narrow your topic as you conduct research (figure 5.1). The scope and focus of your research may change as you start making progress. Simple problems get simple solutions—now you're on to a much more complicated topic: choosing a good mattress.

Figure 5.1. From General to Specific Research Topic. Effective research requires narrowing down to a specific topic to produce a focused search.

General Topic
Sleep Quality

Topic Narrowed by Research
Mattress Studies and Sleep Quality

Narrowed by Additional Research
Comparison of Leading Mattresses

Specific Research Question
Best Mattress for Better Sleep Quality

Step Seven: Locate Credible Sources

If you're reading this textbook because you're in college, you are in luck: a chunk of your tuition pays for access to credible resources and library databases to use for research. In short, you probably have access to independent sleep and mattress studies through the college library's catalog. If Jessamyn still had access her college's database to find recent research on EVs, it would make her research a lot easier. After you've narrowed your topic, you can take advantage of these academic resources to find the studies you need to make your final assessment.

Reference librarians are another resource you can use to find relevant material. Sometimes jokingly referred to as "the original search engine," reference librarians offer much more than a Google search. They can advise you on anything from search techniques to citation methods. They can help you ensure your report on mattresses is the best, most comprehensive mattress report ever.

See your college library's website for more about reference librarians and research assistance.

If you, like Jessamyn, are using a regular internet search, you should limit your search to weed out false results. A basic Google search algorithm privileges paid advertising (usually marked) and so-called relevant search results, which are governed by how frequently the sites are linked or clicked. To get on the top results lists, some businesses buy clicks.[1]

Save time with these three ways to restrict your internet search:

» Do most of your searches using Google Scholar (figure 5.2).
» Use quotation marks around your search phrases to restrict the search to that exact word combination (figure 5.3).
» Restrict your search to .gov or .edu sites to find independently conducted research by using the "site:" function (figure 5.4).

You can also use the "site:" function in a standard Google search to narrow your results to more relevant sources. Even when you take this step, you still need to determine the relevance, accuracy, and credibility of your search results. Don't simply use whatever you find on your first try—dig a little deeper. If you use all these tools to locate sources and come up short, you may need to conduct primary research.

Figure 5.2. Using Google Scholar. Google Scholar compiles resources from academic literature, such as journals, university publishers, and other sites it identifies as scholarly.

Figure 5.3. Conducting Effective Online Searches. The quotation marks tell Google to search for these words as a distinct unit.

Figure 5.4. Locating Educational and Goverment Websties. Websites with a domain of .edu are linked to educational institutions. Websites with a domain of .gov are linked to government institutions. This search limits the sites to those produced by government agencies.

Step Eight: Skim for Relevant Information

When you find a source, evaluate it quickly to determine its relevance or usefulness. Get curious by asking yourself the following questions:

- » Who wrote it?
- » What can you learn about the source?
- » When and where was it published?
- » Does the source provide references?

Take notes as you research so you don't lose the source or your impressions of it as you make your evaluation. If you aren't taking notes, you aren't retaining information. Studies show that handwritten notes help you retain information better than taking notes on your phone or a laptop.[2]

Tips for Skimming

When you skim your document, evaluate the source first. We cover this process of evaluating sources later in this chapter. Where possible, take the following steps to skim documents:

- » Read the abstract, if any.
- » Look for the scope of the research and methods, and review the sources at the end of the document.
- » Read enough of the introduction and the conclusion to determine the reasons and outcomes for the document.
- » Determine whether the source uses language you can't understand, such as jargon or formulas you haven't learned to interpret.

Tips for Taking Notes

There's a saying that the weakest ink is better than the strongest memory. This is why you should take notes whenever you do research. It can save you time and effort to jot down even a few key terms while working through your sources.

Whatever note-taking style you use is up to you, but follow these fundamentals while doing research:

» **Make sure you can find the source again.** Make a bibliography as you find sources instead of waiting until the end of your project. Keep track of important information (author, title, place of publication) that may get lost otherwise. At minimum, write this information down in your research notes.

» **Write down keywords you may need to look up.** You'll need to hit the dictionary a lot more when you use peer-reviewed or scholarly articles. Write down any thoughts you have about the research materials, such as how the source is relevant to your research. This can be part of your screening technique for sorting material quickly.

» **Record useful statistics and data you'll use.** Jot down a quick note reminding yourself why you chose the information.

» **Save page numbers.** This is particularly important if you're using a citation style that requires them.

Don't underestimate the power of note-taking. Effective note-taking requires diligence, but ultimately notes save you time and effort as you compile research for your technical document. You will reference your research often as you compile a report and will be thankful for clearly written notes.

Primary and Secondary Research

Research in technical communication follows two methods: primary and secondary research. For your project to succeed, you need to quickly determine what kind of research is required based on time constraints and available resources. Contrary to the implied order of their names, you will most likely start by skimming secondary research before engaging in primary research.

Primary Research

Primary research is new data that has not been collected before. This is often called fieldwork because it requires running experiments, doing interviews, collecting surveys, and getting out into the world.

If you want to do a research project that develops primary research, you'll need to consider the following:

» **Fieldwork is valuable, but it takes time.** You need to account for how much time you have to complete a project before deciding fieldwork is your best option. Often, effective fieldwork requires exposure to statistics and advanced research methods. If you're unfamiliar with these, leave it to the experts.

» **Effective surveys and interviews provide firsthand experience, but the questions can be difficult to write.** Poorly written survey questions can produce false results. Think about online surveys you encounter periodically. Can you detect that the author wants you to answer in a particular way? If so, that's an invalid survey. If you're interested in conducting an interview or survey, ask your instructor or boss if there are institutional or company guidelines for writing effective questions.

» **Self-reported data is less reliable than other more objective means of data collection.** This is another potential problem you may encounter. If you were doing research on texting while driving among teens as part of a safety study, you would have to account for the possibility that some of the survey responses are probably untrue. This is why you see margins of error listed in polls or surveys conducted by credible sources such as Pew Research Center.

» **Check for existing studies.** Chances are high that the fieldwork you'd like to do has been performed already. Check before you decide to venture out on your own.

An example of fieldwork when choosing a mattress might include conducting stress tests to determine durability, sleeping on the mattress for a hundred days or more to see if you feel more rested, or setting the mattress on fire to check flammability. This takes time and might be expensive. An appropriate time to conduct primary research is when you discover a serious gap in information that will prevent you from making a clear, ethical conclusion.

Secondary Research

Secondary research involves data that already exists. Trade journals, industry publications, and government reports are types of secondary research. This is the kind of research you do most of the time. Before deciding to tackle primary research, look at the secondary research in the field to see if people have already done some (or all) of the research for your topic. Secondary research exists in several levels of credibility and depth (figure 5.5).

Figure 5.5. Hierarchy of Credible Sources. Notice how the bulk of sources are at the bottom of this pyramid. As a researcher, your job is to sort out what's reliable from what's not. And there's a lot that's not.

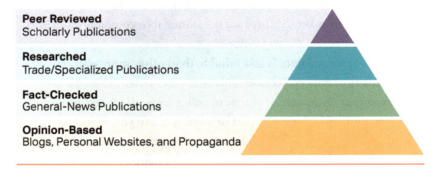

Peer Reviewed
Scholarly Publications

Researched
Trade/Specialized Publications

Fact-Checked
General-News Publications

Opinion-Based
Blogs, Personal Websites, and Propaganda

Level One: Blogs, Personal Websites, and Propaganda

The sources on this level have little to no editorial oversight. Frequently this means that the person writing is responsible for producing, fact-checking, editing, and publishing the content. The source might be a message board or other user-generated site that has few restrictions on content.

You'll need to do some serious background checks if you plan to use sources of this kind. Does the source have an "about" page that defines its stance, publishing standards, or code of ethics? Watch out for sources that have weird URLs, come from personal blogs, or are published on hyper-partisan fringe platforms like U.S. Uncut (on the far left) or Breitbart (on the far right). The chances that you can use any source from this research strata are slim to none, since this level is dominated by opinion, misrepresented data, and amateur reportage.

Level Two: General News Publications

News media is where you usually make first contact with a research topic you'd like to pursue. If you're assigned a project though work or school, general news publications are a good place to get an overview and find the language to use as you dive deeper.

Most news sources have a history of following journalistic standards, such as triple verifying sources before publishing them. Wikipedia articles may fall into this category as a starting place for basic information and useful search terms. These sources are typically not useful for deeper technical reports, but they can put you on a trail, especially if the publications focus on objective analysis and heavily cover science and industry.

Level Three: Trade or Specialized Publications

At this level, you're getting into more credible sources. Trade-oriented publications are accessible to a general audience and a good place to go when you need help understanding the experts.

Examples of specialized publications include *Wired*, *Popular Science*, and *Scientific American*. These publications specialize in general STEM (science, technology, engineering, and math) topics and publish articles that the layperson, or someone outside the specialty, can understand. These publications do long-form journalism or articles that involve substantial research. They usually employ writers with expertise in their subjects.

Level Four: Scholarly Publications

This level features the highest form of credible research because these sources are written by experts in their field and reviewed by a panel of experts for potential flaws prior to publication. This process of examination by a group of experts is called **peer review** and is standard practice in most scholarly works. This level is also where you may find meta-analyses, which are the collected findings of multiple independent research projects into the same or similar subjects to determine statistical significance.

Peer-reviewed journals are tightly focused on specific fields of inquiry, and you can usually find a publication for the field you need to research. For instance, at least six U.S.-based peer-reviewed journals cover fisheries alone, and at least one peer-reviewed journal covers baseball. Scholarly publications are the best resource for deep dives into your research topics.

Using Sources Effectively

One question many students have is how to use sources effectively. The answer: it takes practice. After you have gathered your sources, you need to analyze how this information relates your topic. **Analysis** is the act of examining a complex topic by breaking it down into its specific parts. This examination forms the basis for the discussion of the problem you identified or the solution you are recommending.

In technical communication, unlike academic essay writing, you are less likely to use direct quotations to support your recommendation or solution. Rather, you will summarize, paraphrase, or pull data from your sources. None of these three techniques uses quotation marks, but all require some form of citation.

In the following section, we'll take a look at how Jessamyn summarizes and paraphrases information she found from the University of Michigan's Transportation Research Institute. Figure 5.6 is an excerpt of the source she found.

Summarizing Sources

A **summary** is short description of another source written in your own words. Think of summary writing as similar to how you explain to a friend what a movie is about. It works the same way. After reading an article, you summarize what it said in a paragraph or less to someone who hasn't read it.

Figure 5.6. Report Excerpt. Compare this document with the summary Jessamyn creates from it in figure 5.7. Notice how a summary is significantly shorter than the original source from the University of Michigan's Transportation Research Institute.

Original text for the University of Michigan's Transportation Research Institute.

Introduction

Currently, electricity accounts for just 0.1% of all transportation-related energy consumption in the U.S., while 92% of transportation-related energy consumption is still derived from petroleum (0.03 and 25.7, respectively, out of a total 27.9 quads[1] consumed for transportation) (LLNL/DOE, 2017). However, in recent years, sales of plug-in electric vehicles (PEVs)—both battery electric vehicles (BEVs) and plug-in hybrid electric vehicles (PHEVs)—have begun to accelerate, with sales of each vehicle type increasing by more than 700% since 2011 (AFDC, 2017g). This rapid increase in sales for these relatively new (and still evolving) vehicle technologies was due in part to the need for automobile manufacturers to begin to meet the increasingly stringent requirements to lower CO_2 and other greenhouse gas (GHG) emissions (and the corresponding performance gains in fuel economy) to help comply with current and future CAFE standards.[2] Zero-emission vehicles (ZEVs) such as BEVs have played an important role in recent years to help manufacturers achieve their CAFE targets; California and several other states have recently required the sale of such vehicles (Carley, Duncan, Esposito, Graham, Siddiki, and Zirogiannis, 2016).

Battery electric vehicles (BEVs) operate entirely on electricity stored in on-board battery systems that are charged from the main electrical grid, usually via a special high-voltage charging station and using special electrical connectors. Plug-in hybrid electric vehicles (PHEVs) can also operate on electricity stored in on-board battery systems that are charged from the main electrical grid or by an internal combustion engine (ICE), but with the option of switching to the internal combustion engine for power when the battery runs low. Example illustrations of the key differentiating components for each vehicle type are shown in Figure 1 (AFDC, 2017e, 2017f). The advantage offered by PEVs over conventional ICE vehicles is their ability to operate on little to no petroleum (depending on the vehicle design and operating mode). Correspondingly, little to no CO_2 emissions are associated with such vehicles when calculating CAFE compliance.

[1] One quad (one quadrillion Btu) is equal to approximately 8 billion U.S. gallons of gasoline or 293 billion kWh of electricity.
[2] In March of 2017, the EPA and NHTSA officially announced that the midterm review of CAFE targets for model years 2022-2025 would be re-reviewed (EPA/NHTSA, 2017), reversing the decision to confirm the targets set by the previous administration (EPA, 2017c). Therefore, it is possible that the CAFE targets for 2022-2025 could be altered or eliminated during the upcoming midterm re-review.

1

When you write a summary, you generally include the following information:

» The author's name and the work's title
» The article's main point (conclusion or thesis)
» Key useful details (such as methodology or study size)

Take a look at how Jessamyn summarizes the research she found (figure 5.7). Like her, you should avoid writing long summaries. In early drafts, you might be tempted to fill pages with a "play-by-play" summary of your sources. The research isn't there to provide filler for a report—it's there so you and your audience can better understand your subject and the well-reasoned solution to the problem. Think of the summary as answering the question "what is this source about?" and nothing else.

Figure 5.7. Jessamyn's Summary. An effective summary will contain the original document's most essential information, including the author, main conclusion, and other details that describe the purpose or scope of the research.

The research institute that released the report is considered the author.

This section indicates which vehicles are included in the study.

The summary identifies the conclusion of the original report.

The University of Michigan Transportation Research Institute examines common misnomers of electric vehicles (EVs) in the U.S., including battery-electric vehicles (BEVs), plug-in electric vehicles (PEVs), and plug-in hybrid electric vehicles (PHEVs). The extensive report cites current drawbacks including vehicle costs and accessibility. The research also identifies several benefits of owning EVs. These include less maintenance and fewer overall costs. Because electric energy converts more efficiently than a conventional vehicle, demand is expected to increase. This increase should help lower the purchase price. Once more electric cars are on the road, the U.S. should reap more environmental benefits, such as a reduction in overall emissions compared to gas-powered vehicles and a stronger nationwide security around energy.

Effective Summary

To write an effective summary, you must read the source well enough to be able to explain the information confidently to the audience. A vague, imprecise summary is a sign that the researcher did not thoroughly read or understand the source.

An effective summary gives the audience who is not familiar with the source a clear idea of the article's content. One way to summarize a source is to answer the following questions:

» What is the topic of this research?
» Who conducted the research?
» Why was this research conducted?
» What were the major findings?
» What were the research's requirements or limits?

Paraphrasing Sources

It can be easy to confuse a paraphrase and a summary. A summary describes a source and gives an overview of the material. A **paraphrase** is when you recast an author's specific ideas using your own words. Paraphrased sources have many useful applications in research. For example, you may need to translate technical concepts from a report into language that a general audience can understand. You can use some of the techniques we cover in writing definitions to help you paraphrase.

When you paraphrase a source, it's a little like when you tell a story to someone. You don't remember exactly what someone said, so you relate the general idea to your listener. In Jessamyn's paraphrase in figure 5.8, she translates the report's information into her own words.

Effective Paraphrase

An effective paraphrase requires you to really know your source material. Read the passage you want to paraphrase several times, and then write the passage in your own words. Set aside the original, and don't look at it until you have your own version.

Figure 5.8a. Original Excerpt. Jessamyn begins with the original excerpt.

This is an excerpt from the original report.

In addition to research funding, various government agencies at both the national and state level have enacted legislation specific to PEVs, often with the goal of encouraging or incentivizing vehicle owners (including government, commercial, and individuals) to consider purchasing PEVs.

Figure 5.8b. Paraphrase Based on Original. In addition to paraphrasing the original research, Jessamyn also simplifies the language by avoiding the use of technical jargon ("PEV") to make the paraphrase more understandable for her general audience.

This is Jessamyn's paraphrase of the original source.

Ongoing government support and legislation from both the state and national levels continue to encourage adoption of electric vehicles for individuals, government agencies, and commercial drivers.

If you translate the original word by word or phrase by phrase into your own version, you may end up with a version that is too close to the original. You might be tempted to copy the passage from the original and swap out similar words. But don't. This is not only an example of sloppy research—it is a type of plagiarism.

When you are happy that the paraphrase is in your own words, reread the original to make sure you haven't altered the original idea. Because you borrowed a word, a phrase, a thought, or an idea, you need to cite the paraphrase according to the citation method required by your professor or employer. This allows the user to find the original source for the paraphrased information.

Citing Sources

Throughout college you'll encounter a variety of citation styles. The two styles you are probably most familiar with are Modern Language Association (MLA) and American Psychological Association (APA), but you will likely at some point encounter Chicago style and Institute of Electrical and Electronics Engineers (IEEE) style, if you haven't already.

Citation standards are the agreed-upon system for creating a paper trail for your research. They show other researchers where your information comes from so they can more fully understand your conclusions. Citation also acts as a kind of insurance against claims of plagiarism or copyright violation.

In most instances, general audiences will be satisfied that you use one clear system to identify your sources. Choose the citation system that works best for your audience, and use that system exclusively. You don't want your audience distracted by inconsistencies in your formatting.

In the workplace, no one's going to fault you for citing your sources. Your colleagues and even your future self may thank you for pointing them in the direction of sources they can use for similar projects. When in doubt, use the citation method most common to your industry.

This section is not meant as a thorough review of citation styles. The citation examples that follow demonstrate the main similarities and differences among the most common styles you will encounter as a student. As you review this section, make note of the basic rules, and find an up-to-date reference guide that can help you along the way.

MLA Style

MLA stands for "Modern Language Association," which is the organization founded in 1883 to promote the study of language and literature in the U.S. The MLA citation standards are used by humanities and literature classes in U.S. colleges and universities.

Works Cited

Each works-cited entry in MLA starts with four main pieces of source information:

> » Author name (last name, then first name)
> » Title of source (use quotation marks around shorter works and use italics for longer works like books)
> » Title of container (the name of the publication where the source can be found)
> » Additional information (other contributors, volume and issue number, publisher, date, page numbers for shorter works, and URLs)

A Works Cited page is formatted with a hanging indent and organized alphabetically.

Format

Author. "Title of Source." *Title of Container*, Other contributors (translators or editors), Version (edition), Number (vol. and/or no.), Date, Location (pp.). *Title of Database*, https://doi.org/xx.xxx/yyyy or www.website.com/page/specific-page.html.

Example

Doe, Jane. "Flight of the Killer Alpacas: An Examination of Early Alpaca Literature in Post-Colonial Andean Cultures." *Journal of Historical Alpaca Studies*, vol. 25, no. 3, 2017, pp. 325–415. *JSTOR*, https://doi.org/10.1095/alat/abr073.

In-Text Citation

The in-text (or parenthetical) citation is where you tag the source you've used in the body of your document. The basic expectation for in-text citation is that the author name and page number appear after using the source, as shown in this example:

> "The history of alpacas in post-colonial Andean cultures shows that these hairy creatures are part of a long-standing literary tradition" (Doe 325).

If you have already mentioned the author name in a signal phrase, all you need is the page number in parenthesis at the end, as shown in this example:

> According to Jane Doe, "The history of alpacas in post-colonial Andean cultures shows that these hairy creatures are part of a long-standing literary tradition" (325).

If you have an online source without page numbers, you can use the author name alone (Doe). Notice that the period always goes after the parenthetical citation no matter which kind of source you have.

APA Style

APA stands for "American Psychological Association" and is a style of documentation used primarily by social sciences, such as psychology and sociology.

References

The list of sources for your APA formatted research document is called a reference list. Like MLA, your references are listed in alphabetical order by last name and are formatted with a hanging indent.

Format

Author, A. A., Author, B. B., & Author, C. C. (Year). Title of article. *Title of Periodical, volume number* (issue number), pages, https://doi.org/xx.xxx/yyyy

Example

Doe, J. (2017). Flight of the killer alpacas: An examination of early alpaca literature in post-colonial Andean cultures. *Journal of Historical Alpaca Studies, 25*(3), 325–415, https://doi.org/10.1095/alat/abr073

In-Text Citation

APA style also uses a parenthetical citation like MLA style, but APA citations include the author's last name, the publication year, and the page number each separated by a comma. If you introduce the author's name and publication year in the sentence, however, you do not need a parenthetical citation at the end, as shown in the following example:

> According to Doe (2017), "The history of alpacas in post-colonial Andean cultures shows that these hairy creatures are part of a long-standing literary tradition" (p. 325). (example of direct quote)

If you don't include the author's name and publication within the sentence, then you need to include it at the end as a parenthetical citation.

> Research shows that alpacas play an important role in the literature of post-colonial Andean cultures (Doe, 2017, p. 325). (example of paraphrase)

Page numbers are only required in APA style when you are using a direct quote from a source. However, you may still want to include a page number for paraphrases to show your reader where to find this idea in the original source.

Chicago Style

See *The Chicago Manual of Style* online for more on citations in the author-date style.

The Chicago Manual of Style is a large text that defines **Chicago style**, a type of documentation most often used in the humanities and within the publishing industry. A version of this style called Turabian is often used by students and researchers in the humanities. The Notes-Bibliography system (NB) uses footnotes or endnotes that correspond to sources marked in the text with a raised (superscript) numbers. Chicago style also has an alternate citation system called the author-date style, which is more common in the social sciences.

Bibliography

The list of sources in NB is called the bibliography and organized alphabetically by author.

Format

Author. "Title." *Title of Container* volume number, issue number (publication date): pages, https://doi.org/xx.xxx/yyyy.

Example

Doe, Jane. "Flight of the Killer Alpacas: An Examination of Early Alpaca Literature in Post-Colonial Andean Cultures." *Journal of Historical Alpaca Studies* 25, no. 3 (2017): 325–415, https://doi.org/10.1095/alat/abr073.

Notes

Chicago style uses footnotes or endnotes instead of parenthetical citation. The number appears as a superscript number within the body of the text, as in this sample sentence:

> "The history of alpacas in post-colonial Andean cultures shows that these hairy creatures are part of a long-standing literary tradition."[1]

Footnotes are placed at the bottom of each page. Endnotes are placed at the end of the article or the book. Whether you use footnotes or endnotes, the source information should be documented next to the corresponding number in the list of sources.

Format

1. Author's name (first then last). "Title." *Title of Container* volume number, issue number (publication date): pages, https://doi.org/xx.xxx/yyyy.

Example

1. Jane Doe. "Flight of the Killer Alpacas: An Examination of Early Alpaca Literature in Post-Colonial Andean Cultures." *Journal of Historical Alpaca Studies* 25, no. 3 (2017): 325–415, https://doi.org/10.1095/alat/abr073.

IEEE System

The Institute for Electrical and Electronics Engineers (IEEE) is a professional organization that has created industry standards for the fields of computer science, engineering, and information technology. **IEEE style** is most often used for publications in these areas as well as technical articles and periodicals.

References

References are presented numerically in the order they have been cited in the paper. The author name appears with the first initial followed by the last name.

Format

[1] A. Author, "Title of source," *Title of Container*, volume number, issue number, page range, publication date, doi: xx.xxx/yyyy.

Example

[1] J. Doe, "Flight of the killer alpacas: An examination of early alpaca literature in post-colonial Andean cultures," *Journal of Historical Alpaca Studies*, vol. 25, issue 3, pp. 325-415, Jan. 2017, doi: 10.1095/alat/abr073.

In-Text Citation

IEEE uses a square bracket citation that corresponds to a numbered list of sources on the References page, like this: [1]. You might be familiar with the variant of this system from visiting Wikipedia.

Intellectual Property

By the time you take a technical communication course, you probably know what plagiarism is, and you know that your instructors will probably give you an F for the course if you do it. **Plagiarism** is a serious academic offense that involves presenting another person's ideas, words, or research as your own, which can result in disciplinary action or expulsion from a university.

Plagiarism is not solely an academic issue. Let's say Jessamyn finds a study on electric vehicles and presents it (with a few changes here and there) as her own original research. Even if this action doesn't violate the company's code of conduct, it's likely her boss won't appreciate her dishonesty. But do you know who really won't like it? The researcher who invested her time and research budget to create the original report. During a late-night research session, she googles herself and finds that Jessamyn has posted the research on Tomorrow's Taxi Company's blog. Tomorrow's Taxi Company might be on the receiving end of a cease and desist letter, a demand for them to take down the stolen research or be sued.

As an expert in your field, you need to understand the nuances of intellectual property. **Intellectual property** is defined by the Legal Information Institute at Cornell Law School as "any product of the human intellect that the law protects from unauthorized use by others."[3] For our purposes, we'll touch on copyright, work for hire, fair use and public domain, and Creative Commons.

Copyright

Copyright law governs who owns intellectual property, which can include images and text. The law is simple: if you made it, you own it for the duration of your life, plus seventy years. You don't have to do anything special to show you own your copyrighted material, such as stamp a © symbol on every page of your diary, because the law covers your work as soon as you create it, even if you don't publish it publicly. Copyright law allows designers to seek compensation for their intellectual property.[4]

When someone uploads an image, text, recording, or other content to the internet, that person retains copyright unless they clearly give away those rights. What this means is that you cannot use this content in its entirety without written permission from the creator. The same is true for any content you create.

Work for Hire

If you've entered a **work for hire** contract, which is common in the technical communication field, you don't own what you create for that job. Examples that automatically fall under this contract are workplace communication such as memos, emails, and presentations created as part of your regular job duties. For example, as an employee of Tomorrow's Taxi Company, Jessamyn does not own the rights to her report on electric vehicles. The report belongs to her employer because they paid her to produce it.

If you are involved in a specialized project or working as an independent contractor, you agree in advance to a price for the project and how much copyright the client owns. If you get into a career like graphic or web design, understanding these agreements is vital to your livelihood.

Fair Use and Public Domain

Just because you found it online, doesn't mean it's free. A common mistake people make is assuming they can use whatever they find on the internet. You cannot use copyrighted materials without seeking permission.

U.S. copyright law allows students to use copyrighted works for limited, educational purposes under the "fair use" clause. **Fair use** is narrowly defined as using parts of copyrighted material for specific purposes, such as critique or a short quotation from the original. To use the whole document or major parts of a copyrighted document requires permission from the owner and often involves payment.[5]

When copyright on material ends, it enters the **public domain**. You can use any source in the public domain, which is why so many movie plots are minor variations on Shakespeare plays or fairy tales. Research published through government channels, such as the Centers for Disease Control and Prevention (CDC) or the U.S. Department of Agriculture (USDA), is in the public domain.[6]

Creative Commons

Some people are interested in freely sharing information and apply "Creative Commons" licenses to their work (figure 5.9).[7] This type of licensed material is free to use but may come with conditions. One condition might be

Figure 5.9. Creative Commons Symbol. A Creative Commons license is one of several methods that allows people to use source material without requiring additional permission.

that you can only use the Creative Commons material for noncommercial purposes.

See Chapter 2 for more on Creative Commons licenses.

Whether you use copyrighted material via the fair use clause, a source in the public domain, or material with a Creative Commons license, you should clearly identify and cite your sources.

Advanced Research

Technical communication sometimes requires research skills beyond the academic approaches described in this chapter. If you need to create a document on a topic outside your area of specialty, you may need to work with subject matter experts as sources of information. This section covers how to build communication skills, work with subject matter experts, and conduct usability testing to create more effective technical documents.

Communication Skills

Your primary skill as a technical communicator is the ability to use language in a way that makes your topic easy to understand. Another valuable skill is the ability to learn new topics quickly and relate them to existing knowledge. This is why technical communicators are always in demand. Not everyone has the ability to do this well.

See Chapter 8 for more on the Known-New Contract.

Technical communicators often need to learn about new subject areas quickly. The eight research skills discussed in this chapter provide a good starting place for any topic. Your primary skill is your ability to communicate, not your ability to possess exhaustive knowledge.

Develop an understanding of the difference between key ideas and details or examples. Look for connections between new areas of knowledge and existing knowledge. Often, you'll find that topics about which you know little have a similarity to topics about which you know a lot. Use this to your advantage. The similarities between the known and the unknown can facilitate understanding.

Sometimes, you'll need outside help to learn about a new topic or to check your understanding of a new topic. In these cases, you'll need to look for a subject matter expert.

Consult a Subject Matter Expert

While learning as much as possible about your topic is important, don't assume that you need to be the ultimate authority to create effective technical documents. If you lack mastery of a subject and need to check your accuracy, a **subject matter expert (SME)** can help. Collaborating with an SME is common for technical communicators. You should learn to rely on them, particularly during the research process.

If you embark on a new project in an unfamiliar topic area, find an SME to interview early. Often, your employer will arrange this for you. The SME might be a coworker in a different department of your company. If you are working for a company as part of a temporary contract, the SME might be a full-time employee for that company or another freelancer on a contract. Once the relationship has been established, ask the SME questions to help you in your research. The SME will likely save you considerable time and effort by pointing you in the right direction. Be sure to consult with the SME at the end of your project—their perspective will be invaluable in fact-checking your work.

Don't be intimidated to work with SMEs or to attempt technical writing in new content areas. While most technical communicators have specialized skills and knowledge beyond technical communication, most will be asked to create materials outside of that specialized skill or knowledge. The ability to communicate clearly about complex ideas is a highly desirable skill.

Conduct Usability Testing

See Chapter ⑨ for more on usability testing.

Sometimes part of your research will involve getting feedback from the document's future users through a process called "usability testing." The purpose of this task is to determine whether users understand the document. If they don't, then you need to conduct additional research and revise the document.

Willing and unbiased volunteers to test your document are crucial. If the testers are too familiar with the content or if they have some stake in the content's success, it may be difficult to get reliable and unbiased feedback. Usability testing often requires additional time and is an important part of the research phase of a project.

Case Study

Develop a Research Plan

This case study is an opportunity for you to put into practice what you've learned. Part of this chapter focuses on effective academic and professional research. Look at the following case study to consider how developing a plan before you begin can help you stay on track:

James has been assigned a formal report for his technical communication course. The research process is vital to an effective formal report. James's instructor gave him a choice of topics, so he chose the college's parking situation as his topic. He uses the following research plan and questions to guide his project:

See Chapter 11 for more on formal reports.

- » **Identify the problem:** What is the problem and can it be explained in a single sentence? For example, James wants to examine parking issues on campus, so he needs to identify a specific parking problem. Is there a lack of parking? Is parking too expensive? Are the parking lots inaccessible?
- » **Make observations:** What observations can be made about this problem to formulate a research question that is based on the user's experience? For example, James finds out who is in charge of campus parking and asks questions such as, "How many parking spots are currently available? For students? Faculty? Visitors? How many parking passes are issued per term, per year?"
- » **Develop a hypothesis:** What is an educated guess (hypothesis) about how to solve this problem? Does the solution match the problem? For example, if the problem is that parking is too expensive, then James shouldn't offer a solution that involves adding more parking spots, which would lead to an increase in parking fees.
- » **Test with research:** What does other research say about this hypothesis? For example, James looks into how other colleges handle parking on campus. If he can't find exactly what he's looking for, then his next step will be to investigate how large corporations handle parking.
- » **Consider primary research:** What are the benefits of conducting a survey or interviewing people connected to the problem? For example, James explores whether his college's research and development department can send out a survey about the parking situation to students.

Case Study, continued

» **Evaluate research:** Are the sources credible? For example, James asks himself, "What makes the source credible? Is it relevant? What authority does it have on the topic? And what is the source's purpose?"

» **Record research:** Do research notes identify where to find sources later? For example, James keeps an active list of sources so he doesn't have to compile it the night before the assignment is due.

» **Select research:** Is the report objective and focused? For example, James uses only the material that is relevant to his topic and avoids relying on students' opinions about parking fees. Instead, James focuses on data that shows the increase in parking fees over the past seven years.

Discussion

» Does James's research process look like how you go about research? What's different? What's the same?

» Why do you think James looks for alternate solutions to the parking situation other than the one he is proposing?

» What would you recommend James do next as he begins drafting his formal report?

Checklist for Research

Research Planning

- [ ] Did you make observations about this problem to formulate a research question?
- [ ] Can you explain the problem in a single sentence?
- [ ] Have you made an educated guess (hypothesis) about possible solutions to the problem?

Research Development

- [ ] Have you tested your hypothesis by comparing it with the research you found?
- [ ] What other credible secondary sources did you find?
- [ ] What type of primary research, such as a survey or interview, is needed?

Research Implementation

- [ ] What is the credibility of your sources?
- [ ] Have you summarized or paraphrased research material as needed and cited your sources?
- [ ] Did you confirm that you used only relevant material to your research topic?
- [ ] Is there research material that challenges your hypothesis?

Conclusion

You know how to use Google, but that doesn't mean you know how to research. There's an abundance of information out there, and there's a lot at stake. As a student, you can use your course assignments and research papers to get a handle on how to conduct research effectively and efficiently through library databases and other resources available through your college. If you're good at finding sources, chances are high that college won't be the last time you need to do research.

And if you really love hunting down information, keep developing that skill. Employers are looking for problem solvers. How we find and access information is constantly changing. Employers need people like you who are both creative and systematic in the search for answers or new conclusions.

Notes

1. Kirsten Grind et al., "How Google Interferes with its Search Algorithms and Changes Your Results," *The Wall Street Journal*, November 15, 2019, https://www.wsj.com/articles/how-google-interferes-with-its-search-algorithms-and-changes-your-results-11573823753.
2. Kayla Morehead et al., "Note-Taking Habits of 21st Century College Students: Implications for Student Learning, Memory, and Achievement," *Memory* 27, no. 6 (2019): 807–819, https://doi.org/10.1080/09658211.2019.1569694.
3. "Intellectual Property," Legal Information Institute, Cornell Law School, accessed September 7, 2020, https://www.law.cornell.edu/wex/intellectual_property.
4. U.S. Copyright Office, "Copyright in General," Library of Congress, accessed September 7, 2020, https://www.copyright.gov/help/faq/faq-general.html.
5. U.S. Copyright Office, "Copyright in General," Library of Congress, accessed September 7, 2020, https://www.copyright.gov/fair-use/more-info.html.
6. Copyright Information Center, "Copyright Term and Public Domain in the United States," Cornell University Library, accessed September 7, 2020, https://copyright.cornell.edu/publicdomain.
7. "About CC Licenses," Creative Commons, accessed September 7, 2020, https://creativecommons.org/about/cclicenses/.

Chapter 6

Job Materials

Abstract: Job materials are technical documents that serve as the first point of contact with a potential employer. This chapter introduces the idea of adapting your job materials (your message) to the particular needs of the job (your audience) in order to move to the next stage of the hiring process (your purpose). The fundamentals of technical communication can help you design documents that are persuasive and professional. As you'll see, creating effective job materials is not a one-and-done activity. These materials are living documents that you will build on throughout your career. Like all technical documents, they should be precise, clear, concise, accurate, and scannable.

Looking Ahead

1. Why Job Materials Matter

2. Organize Your Materials

3. Steps for Creating Job Materials

4. Build a Cover Letter

5. Characteristics of Effective Job Materials

6. Job Materials Best Practices

7. Ethical Considerations

8. Be Competitive

Key Terms

» claim
» cliché
» etiquette

» keyword
» references
» S.M.A.R.T.

Why Job Materials Matter

Recent reports from the U.S. Bureau of Labor Statistics show that the average American born in the early 1980s holds an average of 8.2 jobs between the ages of 18 and 32.[1] Younger workers born between 1981 and 1996, known as Millennials, are three times more likely than previous generations to move from job to job, according to a recent Gallup poll.[2] The poll suggests that creating job materials—résumés, cover letters, and the like—is not only a task for recent graduates. Your job materials will likely be a type of technical communication you'll adapt several times throughout your working life.

Almost everyone has spent time looking for a job, so you'd think the average person should be confident in this activity. Not so. Most people feel uneasy when preparing job materials. A survey conducted by Hired, an online employment site, reported that eight in ten working adults find the job search stressful. In fact, the same survey found that most people feel looking for a job is more nerve racking than a root canal.[3]

The job hunt is stressful, true, but it doesn't need to be painful. This chapter deals with the technical skills you need to be successful and addresses some of the myths and misconceptions about the job search. When you create engaging and effective job materials, not only do you increase your chances of getting the job you want, you gain confidence in yourself as well.

The Applicant Situation

Your college writing courses likely taught you how to make a claim and support it with evidence. The documents you provide when applying for a job should do the same. Your **claim**—a statement of what you believe to be true—should be a professional version of "I'm the right person for the job." You support this claim with job materials that show your qualifications.

Your degree, experience, goals, and skills should work together to provide evidence that you are a solid candidate for the position. Fortunately, your coursework in technical communication teaches you to design documents that grab attention and direct the user's eye for maximum impact.

Let's take a look at someone looking for a job right now. Connor will graduate with a bachelor's degree in business administration. He has many high school accolades and spent summers working as a lifeguard. He works part-time in the business office at his uncle's hardware store, but he does not want to work for the family business forever. Recently, he saw a job at Eco-Thrive, a company that builds sustainable tiny homes (figure 6.1). This opportunity could be his dream job, but Connor has some work to do first.

Even with his degree and work experience, Connor is nervous about preparing personal marketing materials. Yes, you read correctly. Job materials should attempt to "sell" your skills to potential employers. Connor needs to decide how to translate his limited work experience and love of the environment into a compelling argument to get him an interview.

Figure 1.1. Job Description. Connor's job search begins with the job posting. Compare the description of this job with Connor's experience.

Business Development Manager

Summary
Eco-Thrive, a leading-edge builder of tiny homes, seeks a Business Development Manager to join its team of dedicated, environmentally conscious employees. The position will oversee daily business operations, maintain accounts, and focus on development and strategic analysis. The ideal candidate will have a degree in business or a related field and know the difference between a flat-head and Phillips screwdriver.

> Whether or not Connor feels ready for a managerial position, he does fit the description.

Qualifications and Skills
- BA or BS degree
- Three years in business sales or related market
- Excellent organizational skills
- Proficiency in Microsoft Word, Excel, PowerPoint
- Superior communication skills, both written and verbal
- Ability to communicate technical information in a clear and concise manner

> The ability to communicate well is vital to all types of companies.

To Apply: Send cover letter, résumé, and three professional references.
Questions? Contact Human Resources at 503-555-5555.

Job Materials

As with other deliverables, job materials are more effective if you focus on your audience, purpose, and message. The fact that your job materials always have the same subject (namely, you) doesn't change the need to tailor the documents to the specific job and audience. No matter how much time you might save by creating a generic set of job materials, don't do it. Human resource managers and the other administrators who review applications typically deal with hundreds of applications for every job opening. If you write generic content for your documents, you will not stand out from the competition.

Your approach to creating job materials impacts your purpose, as well. The way you think about and communicate your purpose should reflect a sincere interest in the position. The generic purpose of "I need a job" is not a compelling argument for your hire. Considering your audience and your purpose together will help you to arrive at a precise message: your job materials. The details of the message should be targeted to the particular audience and purpose as much as possible. You don't need to completely rewrite every job document from scratch, but you do need to avoid sending the same bland document to every potential employer.

Organize Your Materials

Job materials take various forms. For example, a person applying for a job as a nurse or nursing assistant may complete an online or paper application. An applicant for a position as the director of nursing at a hospital may have to prepare a résumé. Why the difference? The first two jobs rely heavily on licensure and experience, while the third also requires the ability to lead, supervise, and communicate well in writing. A résumé allows employers to determine a candidate's abilities in these areas. Figure 6.2 shows a few industry-specific job materials you may encounter.

Figure 1.2. Types of Job Materials. Different job materials serve different purposes. Some jobs may require combinations of these materials.

Job Materials	Industry	Purpose
Application	General	Minimum requirements and specific job-related qualifications
Curriculum vitae	Academic	Comprehensive presentation of qualifications
Résumé	Professional	Concise overview of employment history and qualifications
Portfolio	Creative or Professional	Work samples

Organize by Purpose

Begin organizing your experience by reflecting on your goals. Many career experts recommend starting with goal setting before you create a résumé or cover letter. Why would anyone recommend this? Isn't goal setting an extra, unnecessary step? You don't even have the job yet.

Actually, the process of goal setting is similar to that of research and writing. Think back to what you learned about thesis statements in the past. A strong thesis provides a foundation for your essay or your speech. In the same way, your goal statement can form the basis of your job materials.

Like a thesis statement, your goal statement is a short phrase or sentence that relates to your career objective. If you know what you're aiming for, you can make better decisions and choices along the way. Think of it like a mantra, a repeated phrase that some people use to remind them of what's most important to them. Your goal statement serves a similar purpose in your job search.

To identify your goals, think about what you want. Begin by breaking your goals into current, short-term, and long-term goals. Consider making your goals **S.M.A.R.T.**, an acronym that stands for specific, measurable, attractive, realistic, and time-based. The more you've thought about your goals, the easier it will be to articulate them in writing and in an interview. The strategy of S.M.A.R.T. goals can help you identify what you want in each of the following categories (figure 6.3).

Figure 1.3. The S.M.A.R.T. Technique. Take time to think about what you want, and then write a goal statement to guide your job search.

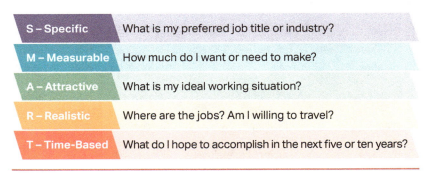

S – Specific	What is my preferred job title or industry?
M – Measurable	How much do I want or need to make?
A – Attractive	What is my ideal working situation?
R – Realistic	Where are the jobs? Am I willing to travel?
T – Time-Based	What do I hope to accomplish in the next five or ten years?

It's important to make a distinction between the goals you set for yourself and the goal statement you might include at the top of a résumé. Your personal goal may be to earn as much money doing as little as possible so you can retire early and live in a windmill. But your goal statement should express not what *you want*, but rather why a specific company would *want you* to be their next hire.

The practice of including a goal statement on a résumé is far from universal, however. In fact, some job experts discourage applicants from including the goal statement because they see these statements as taking up valuable space on the résumé. When in doubt, look for quality résumé examples in your target job market. Do they use a goal statement? If possible, ask the human resources manager at your target place of employment. Do they prefer résumés with goal statements?

Whether or not you include a goal statement in your application packet, the process of articulating who you are and what you want benefits you as you seek positions where you would be a good fit. When you practice articulating your educational experience, relatable job skills, and professional goals, it will help you sound more polished in a job interview.

Select Relevant Details

Be intentional when you choose what details to share in your job materials. Whether you've been working for twenty years or haven't yet held a full-time job, most résumés should stick to a single page. If your experience is deep, you should limit your résumé to the most recent and relevant parts of your work history. If you're new to the job market, organize your experience in other ways by including volunteer work, coursework, skill sets, or activities that relate to the job.

Be selective. Resist the urge to list everything you can think of on your résumé. Your goal is to show that you're an ideal candidate for a specific job, not to overwhelm the person looking at the document with an exhaustive list of unrelated accomplishments.

Create a Distinct Header

Your résumé may be in a folder among many. The header of your résumé is the first data point the hiring manager sees. In fact, many companies are switching to automated services to sort potential hires. In order to stand out, be sure to stay current on relevant resume and hiring trends in your field.

Take a look at Connor's first attempts at his résumé header (figure 6.4). Connor wants to highlight his proficiency with social media, and he has multiple social media accounts that he posts on regularly. Which ones should he include?

Figure 1.4. Résumé Header Variations. Your résumé should stand out in a good way. Consider the effectiveness of these three variations for Connor's header.

Connor Maxwell
Twitter: @onlyonemax — Instagram: @2themax
222 Austen Drive, Portland, OR 97202
Call me at 503-555-1212 or email – tothemax@yahoo.com

- Skip the playful, colorful font.
- Limit social media platforms to those that are relevant to the job.

Connor Maxwell
222 Austen Drive, Portland, OR 97202 503-555-1212
cmaxwell@gmail.com or LinkedIn.com/in/cjmaxwell

- Connor uses a more professional email, but he should avoid hyperlinks, which are unnecessary for a printed document.

CONNOR MAXWELL
222 Austen Drive
Portland, OR 97202
(503) 555-1212
cmaxwell@gmail.com
LinkedIn.com/in/cjmaxwell

- Connor lists only the professional social media platform.
- Connor makes sure his name stands out.

Connor should only include the social media account that is geared toward the workplace and that best markets his skills. It's best if he leaves off the accounts that show where he ate last night or his hobby of knitting sweaters for his pet Yorkie.

Let's take another look at two versions of the header for Connor's résumé (figure 6.5). Notice how the first example gives prime real estate to his career objective. Not only is this objective loaded with general terms that the human resources manager sees several hundred times a day, but it states the obvious. In the second example, Connor ditches the obvious objective, which allows his education and current position to take its place.

Figure 1.5. More Résumé Header Variations. These headers use different approaches to direct the user's eye. The first is a bit crowded, while the second makes use of white space.

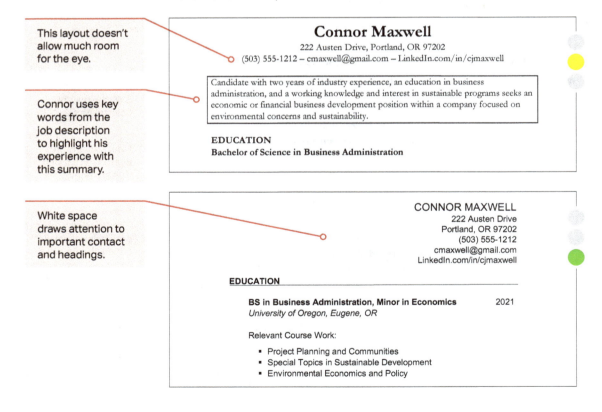

This layout doesn't allow much room for the eye.

Connor uses key words from the job description to highlight his experience with this summary.

Connor Maxwell
222 Austen Drive, Portland, OR 97202
(503) 555-1212 – cmaxwell@gmail.com – LinkedIn.com/in/cjmaxwell

Candidate with two years of industry experience, an education in business administration, and a working knowledge and interest in sustainable programs seeks an economic or financial business development position within a company focused on environmental concerns and sustainability.

EDUCATION
Bachelor of Science in Business Administration

White space draws attention to important contact and headings.

CONNOR MAXWELL
222 Austen Drive
Portland, OR 97202
(503) 555-1212
cmaxwell@gmail.com
LinkedIn.com/in/cjmaxwell

EDUCATION

BS in Business Administration, Minor in Economics   2021
University of Oregon, Eugene, OR

Relevant Course Work:

- Project Planning and Communities
- Special Topics in Sustainable Development
- Environmental Economics and Policy

Organize by Function

A functional résumé design groups similar items together. Your education, work experience, military experience, skills, and so on are all grouped in their own sections, usually beginning with the most relevant or the most recent. Someone with extensive work experience might begin there.

A person right out of college, like Connor, would likely start with his education as the most valuable and then list his work experience. Military experience could be placed in its own section or divided between education and experience depending on the nature of the knowledge or experience gained during enlistment. This list provides some categories to consider when organizing your experience by function (figure 6.6).

Maybe inexperience is not your problem. You volunteer every weekend at the Humane Society, have always had a job if not two, and got a 4.0 GPA in school, a full-ride scholarship, multiple accolades, and employee of the month for four months running. Well, good for you. Your challenge, then, is to decide what to leave off your résumé. Do not reduce the font size so you can cram it all in. Choose the most relevant and recent. You can sum up the depth of your experience in your cover letter and, with luck, in the interview.

Figure 1.6. Organize by Function. These standard categories show how to organize your experience by function.

Education	Experience	Activities	Special Skills
» Related coursework	» Related course projects	» Community involvement	» Social media experience
» Certifications	» Military service	» Volunteer work	» Fluency in another language
» Professional licenses	» Internships	» Leadership experience	» Computer skills

Organize by Theme

A thematic résumé groups items together by categories such as publications, sales, or management—whatever relates to the field. This organization enables experienced individuals to highlight specific aspects of their career successes and areas of knowledge. These thematic categories usually connect directly to key requirements listed on the job posting.

Look for the common threads between your education, work experiences, webinars, and other skills that connect you with the skills the company wants. For example, if the posting mentions market analysis, you might list the jobs where you performed that skill, the professional webinar you completed, and the courses you took on marketing and statistics. On a résumé, these experiences might be listed under the thematic heading "Market Analysis" instead of separately under the headings of education and licensure.

For both approaches, it comes down to preparing the document that best suits the audience's needs. It also matters what information you have to present. The functional approach is the most common and works well for many jobs and many people. The thematic approach can set you apart or organize your document when you have significant experience to relay.

Steps for Creating Job Materials

A plan with a side of research will save you time in the long run and likely result in a more successful job hunt. Start by thinking about your work experience, and then research the potential employer. After taking these steps, you can begin creating your job materials.

Step One: Assess Your Background

Brainstorm all the relevant skills and traits you possess that make you qualified for the job:

» **Experience:** What jobs have you held that show your knowledge in the field and demonstrate a solid work ethic? Did you complete any internships or practicums?

» **Education:** What degrees, professional licenses, or certifications do you hold? Do you have additional college credits or partial degrees? Does any of your coursework show a specific focus in the field?

» **Skills and abilities:** Do you understand how to use computers and various software programs? Are you fluent in a language or languages other than English? Are there field or trade-specific skills you possess?

» **Activities:** Are you part of any professional organizations or clubs? Have you volunteered anywhere that would show knowledge in the field, work ethic, or character?

» **Personal attributes:** What traits make you a good fit for the position? Are you levelheaded? Organized? Good at time management? Do you work well with a team?

Your audience wants to know what makes you qualified, not what makes you interesting. For example, Connor is fluent in Spanish. He highlights his bilingual skills and notes his international travel experience. Connor also likes to hike and considers himself a tea connoisseur. Should he include the latter information? Probably not. Is his experience as a lifeguard relevant? If Connor highlights the skills he acquired during his time at the community pool, it could be. His lifeguard position required an eye for detail, constant vigilance, and calm reactions to potentially life-threatening situations. Those qualities could definitely set him apart.

Step Two: Consider the Employer

Job materials, like the other technical documents, require research, audience awareness, and thoughtful design. If possible, call the Human Resource (HR) department to find out more about the position. In a few minutes, you could learn valuable information about the hiring process for the job. You might gain other valuable information, such as the correct spelling of a manager's name or the closing date for the position.

The hiring process typically has several stages. The first stage of the application might be reviewed by a recruitment manager in the company's HR department. From there, applications may be narrowed and turned over to the department hiring for the position. Your resumé may go through many hands before it reaches the person who interviews you. The more you can learn about a company's hiring process, the more prepared you will be. Does the company do phone interviews or in person? Are you prepared for both? A few minutes of primary research will better prepare you and save you time down the road.

Here are some questions to consider:

» **Research the position:** What are the minimum and preferred qualifications? What are the job duties?
» **Research the company:** How does the company describe itself? Are they local, national, or global?
» **Research the field/industry:** What are the trends in the field? What are the latest industry developments?

Connor looks at Eco-Thrive's job posting and finds that he barely meets the minimum qualifications. He checks out Eco-Thrive's website, reads their mission statement, learns about their recent expansion into California, and finds out that they are a nonprofit supported by government grants. He discovers comparable jobs online to determine what the salary might be. With a little digging, Connor gains valuable knowledge to tailor his job materials and prepare for a possible interview.

Step Three: Prepare Your Materials

From the point you see a job posting that fits your qualifications, you may not have much time to prepare and submit your materials. It pays to think ahead. Connor has taken stock of his relevant qualifications and done his homework. He now has a good feel for the company, the position, and the type of person Eco-Thrive may be looking to hire. The job posting asks applicants to submit a résumé, cover letter, references, and a college transcript online.

The following list provides concrete steps you can take to prepare for a job search:

» Reach out to current and former employers and volunteer organizations who might be contacted during your job search.
» Contact college professors to ask if they would serve as a reference.
» Request official transcripts from all colleges you attended.

Once Connor has completed these preliminary tasks, he begins the process of drafting his materials. As you prepare your own job materials, refer to the three steps in this chapter. Remember the important attributes for all job documents are clarity, simplicity, organization, and concision.

Build a Cover Letter

The cover letter is a strategic document in which you personalize your qualifications. Avoid making your cover letter too long, and someone might actually read it. One page is preferable, but it's always a good idea to find out what's standard in your field.

The cover letter should be memorable and not a restatement of your résumé. For example, Amira is applying for a position at a nursing home. She grew up taking care of her brother who has Down syndrome. The personal experience fuels her professional commitment to providing quality care to others. Amira can't put her personal experience on a résumé, but she can describe the experience in her cover letter.

When assessing your background and choosing the item from your résumé or your life to highlight, you need to select the characteristics that recommend you most. If employers can see you in action in the cover letter, they are one step closer to seeing you do the job.

To begin, think of one defining moment you've had in the workplace. How did it test you? What did you learn? Connor could describe his split-second reaction when he noticed a five-year-old girl struggling to keep her head above water in the community pool's deep end. He could describe how his lifeguard training, his decisiveness, and his attention to the situation allowed him to prevent a potential tragedy. Don't mistake the story for bragging. Connor isn't telling everyone the story so they think he's awesome. He's showing himself in action, and the user is left to conclude that he's attentive and calm under pressure.

As with the résumé, cover letters must be tailored for the specific job application. Does tailoring your cover letter mean you write a completely different cover letter every time you apply for a different job? Not exactly. Certainly, different versions of your cover letter will have similar information, but the specific combination of details will vary depending on the job. You need to mix and match the information you share.

Think of all of the potential details that you could include like toppings in a taco bar. You don't use all of the ingredients every time you make a taco, right? You choose different toppings to keep it interesting. In the same way, don't throw all of your potential information into every cover letter. Sprinkle different information like seasoning into your cover letter according to the particular tastes of the employer. Consider the employer and the job requirements when deciding what to include. To see this approach in action, take a look at the three cover letters Connor has created for this job (figures 6.7 through 6.9).

Figure 1.7. Cover Letter Draft 1. This first letter is polite but offers little information and assumes the résumé speaks for itself. The letter also contains several avoidable mistakes.

Connor Maxwell
222 Austen Drive
Portland, Oregon 97202

December 1, 2021

Taylor Lombard
Director of Human Resources
Eco-Thrive Industries
5555 Greentree Drive, Suite 100
Salem, OR 97301

Dear Mr. Lombardo:

I recently came across Eco-Thrive's posting for an Assistant Business Development Analyst and would like to be considered for the position. My résumé is enclosed for your review. If you need any additional information, please don't hesitate to ask. I can be reached at (503) 555-1212 or tothemax@yahoo.com. Thank you in advance for your time and consideration.

Sincerely,

Connor Maxwell

> Connor has misspelled the director's last name and has also assumed that Taylor is a man.

> The title Connor uses does not match the job posting.

> Connor's email address is unprofessional.

Figure 1.8. Cover Letter Draft 2. The second letter provides more detail but uses large, blocky paragraphs that are hard to read. The letter also contains jargon, typos, and inconsistent formatting.

Connor's formatting is inconsistent. His name and return address should be aligned to the left-hand margin.

Connor has verified the spelling of the HR director's name and has chosen to use the full name instead of making an assumption about the director's gender.

Connor has identified the correct title that Eco-Thrive used.

Connor Maxwell
222 Austen Drive
Portland, Oregon 97202

December 1, 2021

Dear Taylor Lombard:

I recently came across Eco-Thrive's posting for a Business Development Manager on the company's website and would like to be considered for the position. I graduated from the University of Oregon in 2021 with a Bachelor of Science in Business Administration and a 3.6 GPA. I have two years of experience as an Assistant Office Manager for Buck's Hardware and completed an internship for Kelly Motors. In both of these positions, I worked to develop the company's social media presence. At Buck's, I focused on bookkeeping, accounts receivable, accounts payable, and some payroll entry, as well as the preparation of P&L reports. During my internship with Kelly Motors, I had the opportunity to conduct market research and help develop the companies economic strategies and goals for their new line of hybrid vehicles. I would like to venture more into economics with a growing organization and believe your company could help me to expand my knowledge and skills in an environmental area.

Ever since I was a little boy, I have had a strong passion for the environment. I volunteer with the Green Giants, a program focused on promoting reduce, reuse, and recycle programs across Oregon. I have also done work with Homes for Humanity, an organization that builds homes for families in need. Your company sounds like the perfect fit for someone with my interests and I would love to join your team. I look forward to speaking with you at your earliest convenience. If you need any additional information, please do not hesitate to contact me. I can be reached by phone at (555) 555-1212 or by email at cmaxwell@gmail.com. Thank you in advance for your time and consideration.

Sincerely,

Connor Maxwell

Figure 1.9. Cover Letter Draft 3. The final letter demonstrates professionalism through a formal but clear tone that shows effective communication.

December 1, 2021

Taylor Lombard, Director of HR
Eco-Thrive Industries
5555 Greentree Drive, Suite 100
Salem, OR 97301

Dear Taylor Lombard:

I am writing to apply for Eco-Thrive's Business Development Manager position in your community outreach department. I have two years of industry experience, an education in business administration, and a working knowledge and interest in sustainable programs. I believe I can bring a current approach and consistent dedication to the growth of your organization.

My résumé shows that I have put my education to good use even before graduation. An internship with Kelly Motors taught me about economic strategy through performing market analysis studies. At Buck's Hardware, I have gained more technical experience by monitoring the inflow and outflow of cash, taking inventory, and preparing monthly profit and loss reports.

My volunteer work has driven me to focus on environmental sustainability. I served as a team leader with the Green Giants, a local environmental group, for two years. Together, we planted 200 new trees in the Columbia River Valley and helped to clear trails in the Columbia Gorge after devastating wildfires. This experience, combined with my formal work and education, have cultivated a strong work ethic and commitment that I would like to bring to your organization.

I would enjoy the opportunity to hear more about your company in person. If you need any additional information, please do not hesitate to contact me. I can be reached by phone at (555) 555-1212 or by email at cmaxwell@gmail.com. Thank you in advance for your time and consideration.

Sincerely,

Connor Maxwell

Connor correctly includes the name and title of the contact person and the complete address for the company.

Your opening paragraph should focus on what you can offer the company.

Connor refers to his résumé but also offers additional specific information about these experiences.

Characteristics of Effective Job Materials

The principles of designing a user-centered document apply equally here as they do to all forms of technical communication. Presenting your information efficiently makes a positive impression on potential employers. Keep the following principles in mind to ensure your job materials give you the best chance at scoring an interview. Designing job materials is a practical way to put the following principles into practice.

Clarity

In the formatting of all job materials, clarity is king. You need to clearly display your credentials. Consistent formatting helps users find information efficiently (figure 6.10).

Figure 1.10. Alignment in Résumé Design. Use consistent alignment in résumés to create a clean, clear design.

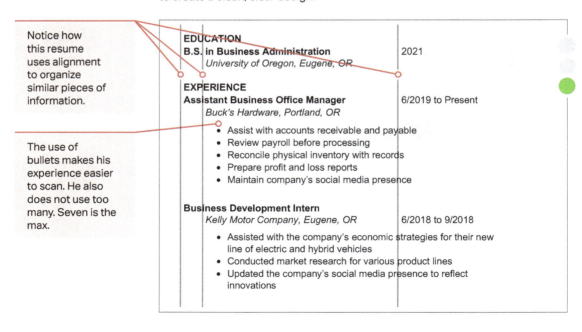

Notice how this resume uses alignment to organize similar pieces of information.

The use of bullets makes his experience easier to scan. He also does not use too many. Seven is the max.

EDUCATION
B.S. in Business Administration   2021
University of Oregon, Eugene, OR

EXPERIENCE
Assistant Business Office Manager   6/2019 to Present
Buck's Hardware, Portland, OR
- Assist with accounts receivable and payable
- Review payroll before processing
- Reconcile physical inventory with records
- Prepare profit and loss reports
- Maintain company's social media presence

Business Development Intern
Kelly Motor Company, Eugene, OR   6/2018 to 9/2018
- Assisted with the company's economic strategies for their new line of electric and hybrid vehicles
- Conducted market research for various product lines
- Updated the company's social media presence to reflect innovations

Professional language does not need to show off. Keep it formal but accessible. Only use trade lingo or field-specific jargon if these terms are useful on job materials. The person doing the recruiting may not understand these terms. For instance, Connor spends time completing P&L reports for his current job. Most people in business understand that P&L stands for "profit and loss." However, the person doing the initial scan of the materials may not be familiar with this term. The recruitment manager may be confused or annoyed rather than impressed.

Simplicity

Clean-looking, standard fonts in one or two colors are preferable for most job materials (figure 6.11). Avoid pictures or images, and keep them to a minimum on web-based portfolios. Occasionally, applicants assume that elaborate page design will attract the attention of potential employers. We don't recommend this strategy. Employers want to be impressed by the content of your materials, not overwhelmed by design features.

Figure 1.11. Simplicity in Résumé Design. Too many colors or fonts can distract from your résumé. Keep the design simple.

Dimitri Stanley

333 West Some Road City, UT 99999 | 801-555-5555 | dstanley@gmail.com

EDUCATION

Degree | Date Earned | School
- Major:
- Minor:
- Related coursework:

SKILLS & ABILITIES

Management
- Offer a summary of management skills from previous jobs.

Communication
- Collaboration is an important skill set; show how you have collaborated in the past.

Keep your design choices simple and easy to read. A consistent serif font and intentional use of color help to differentiate headings.

Employers typically see elaborate design as desperate, rather than creative. Stick to simple, familiar serif fonts that have a professional appearance, such as Times New Romans. The use of sans serif fonts, especially in headings and online résumés, is also acceptable. If you're in doubt, research examples of job documents in your field. Not every job industry has the same standards or expectations when it comes to font choice, so do your homework.

Organization

Intentional organization helps others navigate your job materials quickly to find the information they need.

See Chapter 3 for more on using headings to chunk information.

Begin by thinking carefully about headings in your job materials. Use consistent and deliberate patterns with no more than three heading levels. These headings allow the user to scan the document easily. The content of your headings is equally important, so choose headings that draw attention to themes relevant to the position.

Heading preferences vary depending on the position. For example, if you're preparing a résumé for a job at a nonprofit organization, it would be appropriate to list your previous volunteer work beneath a heading entitled "Volunteer Work." In other jobs, listing volunteer work might be less relevant. Depending on the position, you might skip that heading and dedicate the space to a more appropriate category of experience like licensure.

Bullets and lists are valuable for sharing multiple related ideas that don't require elaboration. Bulleted lists are easy to scan and useful for keeping the user's eye moving down the page. As with all lists, use similar wording for the items that begin the list, and avoid a list longer than seven bullets.

White space groups related information together without fancy separators or text boxes. Be intentional with your use of white space. Effective white space has a few specific benefits:

» It keeps the user's eye moving through the document in a logical pattern.
» It emphasizes the separation and organization of topics.
» It makes the document more attractive by improving visual balance.

Alignment ensures that each element on the page lines up with other similar items on the page, such as dates and work locations. Don't get too creative and make the user work to find the information. Give them a traditional layout that their eyes expect. The key is to use a consistent pattern of alignment so the user easily recognizes the different sections and locates information throughout the document. Notice how these organizational elements come to together in this model (figure 6.12).

Figure 1.12. Résumé Template. This sample résumé provides a pattern that you can use to organize your work experience.

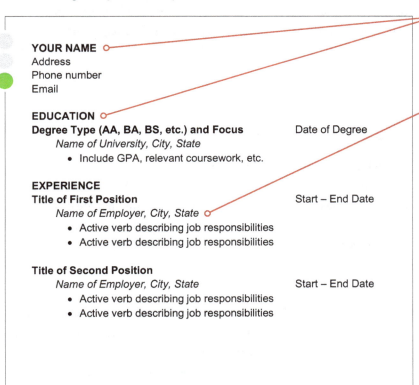

Concision

Most résumés are a single page for a reason. Employers don't have time to read your life story. Make sure your job materials include only what your audience needs to make an informed decision. Do not give them excessive or irrelevant information. Most importantly, avoid including personal data like your gender, race, age, or how many children you have. Employers should not ask for this information, and you should not volunteer it.

Most companies have policies in place to protect you from conscious or unconscious biases that may reduce your chances of getting an interview. But you can help. Leave off hobbies or clubs that don't increase your qualifications for the job. Employers won't be impressed by your knitting skills or Connor's participation in a Dungeons and Dragons group. Anything unrelated to the job could be off-putting.

Job Materials Best Practices

Following a few best practices can make your job materials stand out. Be mindful of how the following elements might impact your audience's response to your materials. Make documents that are easy to scan, include active verbs, avoid clichés, and point to references who can vouch for your work.

Design for Scanning

Research shows that some employers spend as little as six seconds scanning job materials to determine if you might be the right person for the job.[4] Sometimes employers use software to scan job materials for them and identify data points that determine whether the applicant meets the job qualifications (figure 6.13). Once you the attention of an actual person, they are likely to spend five minutes or fewer looking over a résumé. Notice we use the word *looking* and not *reading*. Your document's layout and text must work together to communicate efficiently.

Figure 1.13. Important Résumé Areas. Make good use of the area where the eye tends to concentrate. Place your most recent and relevant experience here.

YOUR NAME

Address | Phone

Email | LinkedIn | Twitter/Blog/Portfolio

If you are including a career objective, keep it short and sweet to avoid taking up prime real estate.

EXPERIENCE

DATES FROM – TO
JOB TITLE, COMPANY

Use this space to concisely describe your responsibilities in this job.

DATES FROM – TO
JOB TITLE, COMPANY

Use this space to concisely describe your responsibilities in this job.

EDUCATION

MONTH YEAR
DEGREE TITLE, SCHOOL

Be proud of your achievements and don't shy away from listing them. This is a good place to mention relevant coursework.

MONTH YEAR
DEGREE TITLE, SCHOOL

Be proud of your achievements and don't shy away from listing them. This is a good place to mention relevant coursework.

SKILLS

· List only skills relevant to the job
· List only skills relevant to the job
· List only skills relevant to the job
· List only skills relevant to the job

ACTIVITIES

Include relevant activities, volunteer experiences, or other leadership roles.

> Research shows that recruiters spend the most time scanning these two areas of a résumé. Place your most recent and relevant experience here.

Most employers spend time scanning for these data points:

» Candidate name
» Current position/company/employment dates
» Previous position/company/employment dates
» Education

Recognizing how your audience will scan your document prepares you to design effective and efficient documents. Your job is to keep the eye moving down the entire page and avoid any design elements or white spaces that stall, misdirect, or stop the user's gaze (figure 6.14).

Figure 1.14. Design for Scanning. This graphic illustrates the way a recruiter's eye flows over the text. Bullet points and white space direct the eye and keep the user moving down the page.

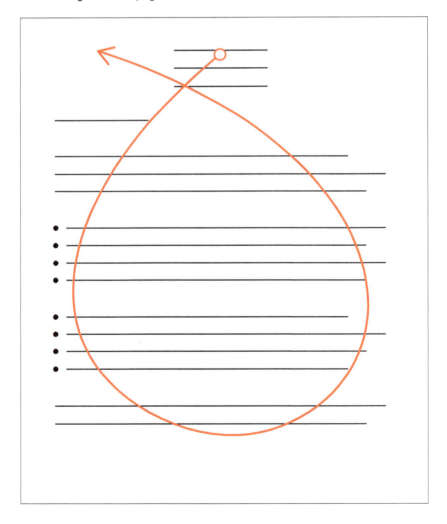

You need to communicate a lot of information in a small amount of space, but extending the margins of your document and creating a wall of text will ensure that your résumé, at best, goes to the bottom of the pile. Too spare of a design can likewise shuffle your résumé to the bottom.

Once a recruiter has determined a candidate possesses the required experience and education, they scan for **keywords,** specific words or phrases that relate directly to the open position. Some companies use software, such as applicant tracking systems, to filter résumés based on keywords. Resume-reading technology is constantly evolving, so it's a good idea to stay up to date on current hiring practices and expectations.

Build your vocabulary list of keywords by completing the following:

» Search the job posting for important words that describe the kind of work required for the position.
» Search professional profiles of people who hold similar positions for how they describe their position.
» Search the company's website and "about us" page for how they describe their work and company culture.

Figures 6.15 and 6.16 show how Connor uses the job posting and the company's website to define keywords for his job materials. In the first example, the highlighted words provide him with a vocabulary list with which to describe his skills and experience on his résumé. In the second example, Connor uses the highlighted words to tailor his cover letter and show how his interests align with the company's mission.

Use Accurate Verbs

You can enliven your technical documents by using more accurate verbs. Precise action words convey the importance of what you've done in previous roles. For example, if you say that you "worked in sales" at a particular job, it doesn't sound particularly impressive. *Worked* can mean many things. On the other hand, if you say that you "managed the sales accounts for forty commercial customers," the user has a much better impression of your role and responsibilities. *Managed* tells the employer that you can handle multiple tasks.

Figure 1.15. Keywords in Job Posting. The highlighted words indicate potential keywords in the job posting.

The highlighted words in this job posting are potential keywords an applicant could use.

Business Development Manager

Summary

Eco-Thrive, a leading-edge builder of tiny homes, seeks a Business Development Manager to join its team of dedicated, environmentally conscious employees. The position will oversee daily business operations, maintain accounts, and focus on development and strategic analysis. The ideal candidate will have a degree in business or a related field and know the difference between a flat-head and Phillips screwdriver.

Qualifications and Skills

- BA or BS degree
- Three years in business sales or related market
- Excellent organizational skills
- Proficiency in Microsoft Word, Excel, PowerPoint
- Superior communication skills, both written and verbal
- Ability to communicate technical information in a clear and concise manner

To Apply: Send cover letter, résumé, and three professional references.
Questions? Contact Human Resources at 503-555-5555.

Figure 1.16. Keywords in Employer's Description. The highlighted words indicate ways to tailor job materials to the specific employer.

The company's website gives clues about their culture. These highlighted words show what the company values in their work and employees.

About Us

Eco-Thrive is a grassroots organization that got its start ten years ago when co-founders Ron and Brian ran into each other at a local hardware store and bemoaned the lack of affordable housing in their community. Combining their love of nature and proficiency with power tools, Eco-Thrive was born. The company has a small but dedicated staff of environmentally conscious employees who are committed to making their community better. Eco-Thrive's mission is to think big but make small by building homes that are affordable and sustainable for all.

On your résumé, use a variety of verbs to convey your past experiences as accurately as possible. Use the past tense, such as "managed," for past jobs. Use the present tense, such as "manage," for jobs you still hold. Specific verbs show the employer that you understand what they want. Figure 6.17 provides a useful list of strong verbs to show your experience.

In your cover letter, go one step farther. Provide evidence of your qualifications that is specific and measurable. For example, Connor volunteered with the Green Giants and helped grow its recycling programs by 20 percent. Generalities, such as "I'm a hard worker" or "I have good leadership skills," tend to fall flat. But statements like "I served as a team leader with the Green Giants for two years" and "We planted 200 new trees in the Columbia River Valley" offer evidence to support his claim. As with any kind of effective writing, a claim should be followed by supporting evidence.

Figure 1.17. Active Verbs for Job Materials. Get more out of the verbs in your job materials. Instead of saying, "I was a manager," say "I managed." It's shorter, more precise, and shows a specific action.

Instead of saying

Experience	Leadership	Problem-Solving	Teamwork

Try saying

Experience	Leadership	Problem-Solving	Teamwork
Coordinated	Developed	Adapted	Assisted
Designed	Chaired	Generated	Collaborated
Organized	Initiated	Improved	Contributed
Maintained	Managed	Increased	Partnered
Specialized	Oversaw	Resolved	Supported

Avoid Clichés

A **cliché** is a sentence or phrase that is overused and shows a lack of originality. How many times can you hear someone tell you to "make lemonade out of lemons" or that they "love you to the moon and back?" It might be cute, but being cute is not the skill you're marketing.

You might think employers want to hear that you're a leader, excellent at interpersonal communication, and detail oriented, but everyone uses these phrases. Do your research and find the overused phrases in the specific job document you're completing. Find a different way to convey that information without employing the clichés.

What you need, instead, are keywords. As mentioned earlier, these words are functional rather than descriptive. An example of functional keywords would be to say you are a "social media expert" instead of saying you have "excellent communication skills." Employers use keywords to find matches, particularly when your materials are part of a database. Saying you are a "strong leader" will lump you in with a ton of other people. This table outlines phrases overused by job seekers (figure 6.18).

Figure 1.18. Effective Word Choice for Job Materials. If you want to stand out, don't use tired phrases and buzz words.

Clichés to Avoid	Attention-Grabbing Alternatives
Strong leadership skills	Led department to a 30 percent increase in client satisfaction
Good communicator	Excellent at generating approved grant proposals
Excellent team player	Worked with a team to increase company revenues by 10 percent
Self-motivated	Independently implemented a new customer satisfaction program
Track record of success	Consistently surpassed department goals by 5–10 percent

Include specifics when possible as further evidence of your claim. Don't underestimate the power of details. Anyone can describe themselves using adjectives, but competitive applicants can point to what they've done to demonstrate their character and abilities. Even without data to measure your success, specific language builds more credibility with your audience. It also proves that you're a "good communicator" without you having to say it.

Maintain Strong Connections

Job postings generally ask for three **references** or recommendations, either as part of the application process or part of the interview. It's important that you don't wait until you're looking for a job to build your bench of supporters who are familiar with your work ethic and abilities. Keep in touch and stay on good terms with previous supervisors and teachers. These contacts are most likely to be able to provide a recommendation.

Connor has already asked two of his current instructors to serve as references. He considered using his uncle as a reference, since Uncle Buck is his current employer and direct supervisor. However, a family member isn't usually the best, unbiased choice. Besides, Connor's uncle doesn't know he is job hunting. So, what should he do?

Most potential employers do not expect you to use your current employer as a reference. Many people look for a new job while they still have one, often without telling their boss. You do not need to jeopardize your current position by sharing this information. Connor can leave his uncle off his reference list without alarming anyone. Instead, he contacts the coach of his swim team, who can speak to his commitment to continuous improvement and his ability to work well with others.

Ethical Considerations

See Chapter **2** for more on ethics in the workplace.

Tailoring your job materials does not mean exaggerating your abilities or past jobs. It means sending the right materials for the right position, not altering history. Even simple adjustments, like referring to your host position at a restaurant as a "client services manager" distorts the truth and implies an unproven level of skill. Your dishonesty will eventually be exposed when it is time to do the work.

Connor does not supervise anyone in his current job, but he knows leadership skills will make his résumé stand out. Rather than counting the times he has kept his uncle's children from destroying the office as supervisory experience, he highlights his ability to work as part of a team and his leadership within his volunteer groups.

Be Competitive

Let's take a look at how Connor pulls together his final résumé. He makes several attempts to strike the right note, but after some research and revision, Connor is prepared to make a good case for himself.

Look closely at how these documents change and how you can apply these principles to job materials and other technical documents you may create (figures 6.19 and 6.20).

Figure 1.19. Draft Résumé. Think about how you could incorporate consistent design features to improve this résumé.

Connor Maxwell

222 Austen Drive, Portland, OR 97202
(503) 555-1212 – cmaxwell@gmail.com – LinkedIn.com/in/cjmaxwell

Candidate with good customer service, interpersonal, and communication skills seeks employment in a growing company where I can make use of my education and work experience and grow my skills.

EDUCATION
Bachelor of Science in Business Administration – University of Oregon, Eugene Minor – Economics
Grade Point Average – 3.6 out of possible 4.0

EXPERIENCE
Assistant Business Office Manager – Buck's Hardware, Portland, OR June 2019 to Present
Business Development Intern – Kelly Motor Company, Eugene, OR June 2018 to September 2018
Team Leader – Volunteer – Green Giants, Eugene, OR
September 2016 to Present
COMMUNITY ACTIVITIES & AFFILIATIONS
Volunteer for Homes for Humanity – National organization building homes for people in need   2019 to Present
Team Leader for the Green Giants – Local environmental group
            2016 to Present
Member of the Future Leaders of America – Oregon Chapter focused on developing youth leadership   2014 to 2017

Connor's name definitely stands out, but at the expense of detail later on.

This career objective does not provide specific key terms from the job description and takes up valuable space on the résumé.

Irregular formatting looks messy and unprofessional.

Figure 1.20. Final Résumé. It may take a few attempts, but a solid résumé is worth the effort.

Connor can make his name and the categories of his resume stand out by placing them in all caps.

Right-aligned text draws the user's eye to important information.

Consistent indents make it easier for the user to track the items in this list.

Bullet points use parallel verbs to describe Connor's work duties.

CONNOR MAXWELL
222 Austen Drive
Portland, OR 97202
(503) 555-1212
cmaxwell@gmail.com
LinkedIn.com/in/cjmaxwell

EDUCATION

BS in Business Administration, Minor in Economics   2021
University of Oregon, Eugene, OR

Relevant Course Work:

- Project Planning and Communities
- Special Topics in Sustainable Development
- Environmental Economics and Policy

EXPERIENCE

Assistant Business Office Manager   June 2019–Present
Buck's Hardware, Portland, OR

- Assist with accounts receivable and payable
- Review payroll before processing
- Reconcile physical inventory with records
- Prepare profit and loss reports
- Maintain company's social media presence

Business Development Intern   June 2018–Sept. 2018
Kelly Motor Company, Eugene, OR

- Assisted with economic strategies for the company's new line of electric and hybrid vehicles

Figure 6.20. Final Résumé Continued. It may take a few attempts, but a solid résumé is worth the effort.

- Updated the company's social media presence to reflect innovations
- Conducted market research for various product lines

Team Leader, Volunteer   Sept. 2016–Present
Green Giants, Eugene, OR

- Promoted and helped update recycling programs in three school districts
- Led a volunteer team on a tree-planting project in the Columbia River Valley in 2016

COMMUNITY ACTIVITIES & AFFILIATIONS

Volunteer for Homes for Humanity   2019–Present
National organization building homes for people in need

Team Leader for the Green Giants   2016–Present
Local environmental group

Member of the Future Leaders of America   2014–2017
Oregon Chapter focused on developing youth leadership

SKILLS

Computer: Proficient with Microsoft Word, Excel, and PowerPoint
Social Media: Experienced with Facebook, Twitter, and Instagram
Languages: Fluent in Spanish

Just because he didn't get paid to do this work doesn't mean it's not important to a future employer. It belongs under "Experience."

Each section handles lists differently, but each is internally consistent.

Business Etiquette

Every career field has a certain level of **etiquette**, an agreed upon set of behaviors. You may not know what is typical in your field yet, but there are general guidelines to follow.

» **Use standard written English:** Your written communication should always consider your audience and be proofread. Even when a potential employer seems a bit more relaxed, maintain a professional tone. Avoid slang, abbreviations, shortcuts (LOL), politically incorrect terms, and emoji. You never know when you are being reviewed.

» **Be polite:** You may not like that the HR manager does not respond promptly to your calls or emails or is a bit of a know-it-all. Right now, your focus should be on getting the job. Brush it off and always use please and thank you.

» **Don't go overboard:** Too many colors or fancy fonts, scented or colored paper, and unnecessary images or graphics can be tacky and make your documents hard to read. You want your materials to stand out for the right reasons.

» **Avoid assumptions:** If you don't know how someone identifies, don't assume. Don't assume the hiring manager is male. Additionally, don't assume all females are married. To avoid making a misstep, you can use their full name instead.

» **Follow up:** If you receive an interview, send a brief thank-you note or email immediately after the interview. Like all job materials, this communication should be targeted. Be specific and thank the person by name who interviewed you. If multiple people interviewed you, then send thank-you notes or emails to each.

Case Study

Business Etiquette

This case study is an opportunity for you to put into practice what you've learned. Part of this chapter focuses on developing professional-looking job materials and understanding how you build your reputation every step of the way. Look at the following case study to consider how etiquette and ethics relate to each other within the context of a job search:

While Connor has his eye on the position at Eco-Thrive, he decided to apply to a position at another company. He was offered an interview at a company called In Dive, which he accepted.

However, after the interview, Connor realizes that the job at In Dive is not at all what he wants. Should he wait to see if they offer him the job before he declines, or should he write a thank-you email after the interview and let them know he has decided the job is not for him? He wants to be respectful of everyone's time. After all, he was impressed with the company, but felt the position didn't fit his goals. Take a look at Connor's email drafts to see which one you think he should send (figures 6.21a and 6.21b).

Figure 6.21a. Thank-You Email. In this first draft of his email, Connor thanks his interviewers and keeps his options open.

Figure 6.21b. No Thank-You Email. In this first draft of his email, Connor thanks his interviewers and explains that the job is not for him.

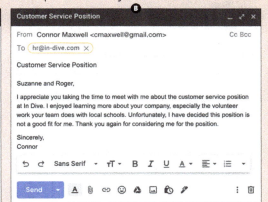

Discussion

» Which email would you send? Why?

» What changes would you make? Why?

» Write an email that expresses interest in the job at In Dive. Consider tone, clarity, and organization.

Checklist for Job Searches

Planning

- ☐ Have you researched the company or organization to learn more about them?
- ☐ Have you looked into what type of résumé is most effective for the field?
- ☐ Do you have a list of key terms to use?
- ☐ Are there any other requested documents to submit with your résumé?

Writing

- ☐ Have you addressed the cover letter to the correct person?
- ☐ What organizational pattern will work best for your résumé?
- ☐ Are your headings, fonts, and layout consistent?
- ☐ Did you proofread your cover letter and résumé?

Follow Up

- ☐ Have you written a thank-you letter or email to the hiring manager?
- ☐ Do you know the next steps in the hiring process?
- ☐ Are your references ready to send, if requested?

Conclusion

The fact that so many people find applying for a job stressful can work to your advantage. If you are mindful of the recommendations in this chapter and thoughtfully create your job documents, you can set yourself apart from the competition. Producing clear and engaging job materials is not as difficult as many people think. Doing so requires clarity and an awareness of your audience, as with all technical communication.

Your job materials are strategic marketing tools. You may feel uncomfortable with that idea, but consider how many other people may be applying for the same position. You must strategically promote yourself and your skills in order to stand out.

Remember that most colleges and community centers offer job material workshops. Take advantage of these resources. Your qualifications will not speak for themselves. Your document must do the talking and build the claim that you are the right person for the job.

And what if you're not yet the right person for the job? Take this experience as your clue to start looking for opportunities, courses, mentors, and internships to build the skills employers seek.

Notes

1. U.S. Bureau of Labor Statistics, "Americans at Age 33: Labor Market Activity, Education and Partner Status Summary," Released May 5, 2020, https://www.bls.gov/news.release /nlsyth.nr0.htm.

2. Amy Adkins, "Millennials: The Job-Hopping Generation," *Gallup* online, accessed August 27, 2020, https://www.gallup.com/workplace/231587/millennials-job-hopping-generation.aspx.

3. Robin Madell, "What's More Stressful than a Root Canal? Finding a Job," *U.S. News and World Report,* December 12, 2016, accessed September 8, 2020, https://money.usnews.com/money/blogs /outside-voices-careers/articles/2016-12-12/whats-more-stressful-than-a-root-canal-finding-a-job.

4. "Getting Your Resume Read," *Deloitte* online, accessed August 27, 2020, https://www2.deloitte .com/us/en/pages/careers/articles/join-deloitte-recruiting-tips-getting-your-resume-read.html.

Chapter 7

Workplace Communication

Abstract: This chapter explores the purpose, types, etiquette, and ethics of workplace communication. The documents we use in daily workplace communication are an essential part of technical communication and require attention to audience, purpose, and message. This chapter covers the conventions, format, and style of different types of workplace communication, including emails, video and phone calls, memos, minutes, agendas, business letters, and even text messages. Technology continues to change how we communicate in the workplace. However, the fundamentals of technical communication remain the same and can prepare you to succeed in the future.

Looking Ahead

1. Why Workplace Communication Matters
2. Types of Workplace Communication
3. Communicating Professionally
4. Collaboration
5. Global Workplace Etiquette

Key Terms

- » agenda
- » block format
- » collaboration
- » directive
- » indented format
- » memo
- » minutes
- » modified block format
- » thread
- » tone
- » workplace communication

Why Workplace Communication Matters

You might not think of a business letter or memo as technical communication. But they are. Remember that technical communication means delivering specific and precise information to users. In the workplace, both the message you send and the people who read it have a unique job. As a result, workplace communication requires a technical approach.

Take the humble email, for example. Frank, an executive assistant, spends most of his day sending emails. When his supervisor asks him to send vital project information to his team, Frank finds a group email from a few weeks ago and hits "reply all." He cuts and pastes tables from the project report into the email and sends it out. What do you think happens?

You guessed it. Most don't read the email because the subject line appears unrelated to the recent meeting. Even those who do open the email aren't sure what they're to do with the information they received. Frank hasn't saved any time, for himself or anyone else.

An effective communicator makes the email's purpose clear by starting a new thread and giving it a timely subject line. The email opens with the most important information. The data tables include a heading and explanation of the content. It then closes with a specific request for action or response.

To communicate effectively in the workplace, you need to keep in mind the audience's expectations for content and format. You're going to spend most of your career communicating with others at work, either face-to-face or virtually. You can distinguish yourself and create an efficient workflow by learning the expectations and standards of workplace communication.

Workplace Communication Defined

Workplace communication is any information sent from one person to another to accomplish a task. This includes staff meetings, emails, memos, business letters, and more. You've certainly written emails to your family, led a scout meeting, or sent a text to your landlord about the leaky roof.

But the workplace equivalent of these communication tools requires a technical approach.

As with other types of technical communication, workplace communication seeks to solve a problem, and this problem takes the form of a need. Workplace needs can be sorted into the following categories:

» A need for information
» A need for instruction
» A need for persuasion

For example, your coworker missed a meeting due to illness. They need information, so you share your notes in the form of meeting minutes. Alternatively, you're convinced that a policy in the employee handbook needs to be updated, so you send a proposal by email to persuade the HR manager to reconsider the issue. The success of these workplace documents depends on considering your audience.

When you're working with others, be sure to remember that they're people. They have strengths and weaknesses and families and problems—just like you. Consider how your words will affect others. Realize that people have limited time and limited attention spans. Keeping your document tightly focused on your specific goal will make your communication more likely to be successful.

Knowing who you're writing to (your audience) and why you're writing (your purpose) should guide your message and direct you to the most appropriate medium (an email, a letter, an agenda). Is your audience the executive board or your workplace bowling team? Should your message be formal or informal? Answering these questions will get you moving in the right direction.

See Chapter **1** for more on purpose, message, and audience.

Types of Workplace Communication

Preferred modes of workplace communication have changed rapidly in recent years. According to a 2020 Gallup survey, over half the world's population now has internet access. In the U.S., 94% of adults reported having internet access in the past year.[1] The reliance on digital communication tools continues to grow both in the workplace and in other sectors of our lives, including education and healthcare. As a result, you need to be familiar with a broad range of communication tools to be able to function in the professional world. You need to know the advantages of these tools as well as how to use them appropriately to get work done.

Email

Despite all the spam and surveys flooding your inbox, email is still one of the most common forms of workplace communication. If you're not careful, however, an email can easily become unprofessional.

Have you ever sent an email to an instructor or potential employer and never received a reply? After reading this section, take another look at your sent folder to see what you could've done differently.

Professional Use

Email is useful for short messages of about one to three brief paragraphs. Email is also an effective tool for sending links and files, such as Word documents or PDFs. If you want the recipient to read your email, one strategy is to use the content window in your email program to limit your message. The goal is to write messages that do not require scrolling. If you must scroll up and down too much to read over your email, your user is unlikely to read the whole message. As with most technical communication, begin with the most important information first.

Email dialogues are called **threads**. The sender starts the thread, and recipients reply to it, creating a chain of responses. This groups related replies under the same subject line. If you want to email someone about a new topic, you should always start a new thread.

Formatting

Business emails are more formal than personal emails, so even if you're sending a quick note to a team member, be aware of formatting and how it affects readability.

In the following sample emails, compare the message between friends who are coworkers and the message between Leticia and her boss (figures 7.1 and 7.2). What do you notice about how the emails open, how Leticia adjusts her tone in each, and how she closes the emails?

You can increase the readability of your email by using shorter block paragraphs separated by white space, as shown in Leticia's email. Providing a subject line is another way to make sure your message registers as important. If the subject line reads "Info: Report on development progress," the user knows that this is an update with no need to respond. If the subject reads "Urgent: Investor meeting this afternoon," the audience knows they should open the email and review it right away.

See Chapter 3 for more on readability and white space.

If you need to discuss multiple topics, consider including a bullet-point summary at the beginning of your email and use headings to clearly indicate the sections in which each bullet is discussed. This is more common in company-wide emails sent to multiple recipients or updates that can be scanned quickly for relevant information.

Figure 7.1. Informal Workplace Email. These colleagues have been working together for a long time on this project, so they've dispensed with some of the formalities expected in a workplace email.

> **From:** Jason A. Mendoza <jamendoza@gamingcentral.com>
> **To:** Leticia White <lwhite@gamingcentral.com>
> **Subject:** Character profiles for review
> **Attachment:** characterprofiledraft.doc
>
> Hey LW —
>
> I have some of the first drafts of character profiles for *Alpacas!* See attached. Feedback?
>
> JAM

Jason knows his audience, so this level of informality is acceptable.

Figure 7.2. Formal Workplace Email. The email is brief, to the point, and contains all the information the boss needs to respond to the request.

Leticia is the main point of contact, so she addresses the email to her boss and copies Jason using the CC: field.

Leticia's subject line makes it clear what she needs from her boss.

Mention attachments — some business email systems may not display attachments.

From: Leticia White <lwhite@gamingcentral.com>
To: Robert Ossman <bossman@gamingcentral.com>
CC: Jason A. Mendoza <jamendoza@gamingcentral.com>
BC:
Subject: For Review: Alpacas of Doom proposal
Attachment: Alpacas_proposal_final.doc

Dear Mr. Ossman:

My creative partner Jason Mendoza and I have completed the *Alpacas of Doom!* game proposal for your review. Please see the attached file. It includes an overview of the game, gameplay, basic and advanced features, and an update on the development process.

We believe the target audience is ready for a game featuring these furry Peruvian villains. Jason and I are available to meet with you in person or by web conference to answer any additional questions you might have about the game. We are excited to hear your feedback on our proposal.

Sincerely,

Leticia White
Senior Game Designer
Gaming Central, Inc.
lwhite@gamingcentral.com
(541) 555-5555

Phone and Video Calls

If you find yourself going back and forth by instant message or email, you may save time by picking up the phone or requesting a video call. In recent years, more people have begun working remotely, and with this change comes an increased use of video calls and virtual meetings.

Whether an email chain has gone too long or you're working from a home office or hotel room, you can often get more information and context from a phone or video conversation than an email or text thread. Context clues, such as tone of voice or facial expressions, can reduce misunderstanding. A real-time conversation also gives you a chance to ask questions and receive answers right away.

Professional Use

Phone and video calls are like texts without the permanent record. If someone is recording, that's another matter. That should be made clear at the beginning of the call. If the call is not recorded, you can use it to discuss sensitive project information, brainstorm complicated subjects, and outline plans before committing anything to a permanent file or database. This format is also a better option for working through complex issues with colleagues or clients.

As a form of communication, phone and video calls are multimodal and incorporate aural, visual, gestural, and linguistic modes in real time to help the user better understand the message.

Formatting a Phone Call

Formatting a phone is similar to writing an email with a salutation, body text, and closing. In general, you greet the caller and say who you are and why you're calling. When speaking to a client, an employer, or a colleague, you might schedule the call in advance to make sure the person has time to speak. Sometimes there's small talk if you know the person, but, for the most part, you should get to the subject of the call quickly.

If you make a phone call and don't reach the individual, you'll normally be greeted by a recording asking you to leave a message. Consider this voicemail message as a document that you're leaving behind. Like all the documents discussed in this textbook, your message should communicate effectively.

See Chapter 4 for more on multimodal communication.

Here's an example of what not to do: "Hey, just returning your call. Call me back." This caller assumes that the person on the receiving end recognizes their voice. Don't assume that the caller ID will provide your name and callback number. Instead, try this: "Hello, Ron. This is Stephanie. I'm calling about the agenda for tomorrow's meeting. I'll be at my desk for the rest of the day. Please call me back at extension 1234 when you get this message. Bye."

The voicemail identifies the caller, explains the purpose of the call, and provides call-back information and a time when the caller can be reached.

Formatting a Video Call

Many types of software exist for video calls, and most allow the user to format a meeting. Enter a subject that is straightforward and one that participants will recognize. For example, use a specific subject like "Budget Planning Meeting." Video calls are generally scheduled in advance to make sure all parties can attend. If your team is working in different time zones, be sensitive to calls that might be too early or late for some participants.

Video call etiquette can vary, but some basic responsibilities include the following:

» Dress professionally.
» Show up on time.
» Limit distractions.
» Prepare for the meeting.

If possible, use a computer with a camera and leave it on or off during the meeting depending on the expectation of the work group. Or at least turn it on when you speak. Limit background noises and use mute when you're not speaking. Use a neutral place in your home or office or use a background provided by the application. Even though you're at your computer, resist the urge to multitask during the meeting.

Video calls are multimodal and usually provide more than one way to interact. Using common workplace etiquette and being aware of new technologies allows people in different places to communicate effectively.

Memos

The **memo**, short for "memorandum," is a brief document distributed in hard copy or in an email attachment. Often a memo is shared internally at a workplace or within another closed group.

Professional Use

The purpose of a memo is to communicate briefly to a specific audience. The information should be easily and quickly understood. However you choose to format the memo, its subject should not be too complicated. A change in parking protocol, a policy change announcement, or a reminder to remove rotting takeout food from the shared refrigerator on Fridays are appropriate subjects for a memo. Using a memo to announce the decision to lay off all part-time employees, on the other hand, would not be a wise use of this document.

Whatever your subject, the language in a memo should be direct and clear. As with any professional communication, memos are an efficient way to communicate simple messages. They are not ideal for navigating complex problems or dealing with topics that may be emotional in nature. For example, a **directive** is a type of memo that issues an order to staff. If the order is complex, the memo should include an opportunity for a meeting or a procedure for recipients to express concerns and ask questions.

Formatting

Your workplace may use its own letterhead for memos. Otherwise, the formatting is simple. The word "Memo" or "Memorandum" is centered or left aligned at the top of the page. Next comes a header, also placed at the top left of the page, that includes the recipient, sender, date, and subject.

The memo is the predecessor of the email. In fact, many memos are now sent as emails. Like an email, the body of a memo should focus on one topic. It will likely include a brief introduction, followed by one or two short paragraphs, and conclude with a call to action. Memos have a standard, consistent format that includes the elements shown in this model (figure 7.3).

Figure 7.3. Internal Memo Example. Memos are often printed on letterhead because they are official forms of business communication. Memos open with these elements: To, From, Date, and Subject.

Printed memos are often signed with the initials of the authorizing party.

This memo begins with a summary and breaks down information using headings for ease of reading.

This is just the first page of a two-page memo. Patsy will email this out to all employees as well as place a printed copy in their mailboxes.

HERRERA, LINDSAY & ORLOV

Memorandum

To: All Employees
From: Patricia Good, Assistant Manager *PG*

Date: November 1, 2019
Subject: International Client Communication

Herrera, Lindsay, & Orlov is excited to announce the acquisition of several new international accounts. This expansion brings the need for updated policies covering client communication to promote healthy working relationships. Several informational sessions and trainings will be offered before putting new policies in place.

Informational Sessions

Informational sessions will be held beginning Friday, November 2, from 3:30 to 4:30 p.m. in the staff lounge to discuss policy changes.

Trainings

- International Client Communication
  December 1, 9 a.m. to noon
- Cultural Competency and Client Relations
  December 15, noon to 5 p.m.

Timeline

After collecting employee feedback during the informational sessions, we will establish a committee to revise the current policy.

4001 LANCASTER DR NE, SALEM, OR 97309
T 541-555-5555 U WWW.HLO.COM

Meeting Agendas

Sometimes, the best way to adequately convey a message, brainstorm a project, or consume birthday cupcakes is in a standard face-to-face meeting. You can make sure you're all on the same page because everyone is in the same room holding a copy of that page, a document that defines the purpose of the meeting.

Professional Use

Efficient meetings make use of a guiding document that is preferably distributed ahead of time. The **agenda** is a document that outlines the topics of discussion for a meeting. Think of an agenda like a table of contents for the next hour or so. A meeting without an agenda can easily get off topic.

Formatting

If you need to hold a meeting, make sure the agenda states the purpose, topics for discussion, and the order in which presenters will take turns speaking. An agenda, like the one shown in figure 7.4, will provide an outline for the meeting.

Meeting Minutes

Meetings are most often led by one person while another takes notes called **minutes** during the meeting. These notes are usually distributed later to participants and people who couldn't make it to the actual meeting. Businesses often keep minutes as a permanent record of decisions, sometimes by legal requirement. Companies usually have a basic meeting format to follow, as shown in figure 7.5.

Professional Use

Meeting minutes are not a detailed description of the meeting itself but instead a record of the meeting's outcome. If people disagree with one another during a meeting but reach a conclusion, the disagreement is not included in the minutes, only the result.

Figure 7.4. Meeting Agenda Example. The agenda creates an outline and plan for the meeting.

HERRERA, LINDSAY & ORLOV

Cultural Competency Committee Meeting Agenda

Location: Orlov Board Room
Date: November 15, 2019
Time: 10 a.m.

Welcome

Approval of Minutes

Vote on Chair for the Cultural Competency Committee

Report on Training Sessions

- International Client Communication
  December 1, 9 a.m. to noon
- Cultural Competency and Client Relations
  December 15, noon to 5 p.m.

New Business

- Review of employee feedback
- Draft of new client communication policy

Next Meeting

- Decide on date

Meeting Close

4001 LANCASTER DR NE, SALEM, OR 97309
T 541-555-5555 U WWW.HLO.COM

Figure 7.5. Meeting Minutes Example. Minutes record the decisions made during the meeting.

HERRERA, LINDSAY & ORLOV

Cultural Competency Committee Meeting Minutes

Location: Orlov Board Room
Date: November 15, 2021
Time: 10–11 a.m.

Attendance: Patricia Good, Jaime Herrera, Ralph Lindsay, Jodie Orlov, Astrid Middlestadt, Richard Carlin, Matthias Chekov

Item #1 – Vote on Chair

- Patricia was unanimously voted to be the chair for the Cultural Competency Committee. She will serve a two-year term.

Item #2 – Report on Training Sessions

- Both training sessions have open spots. Astrid will send out a company-wide email by November 21 to encourage signups.

Item #3 – Review of Employee Feedback

- Employee response rate to the survey sent November 1 is 45 percent. Matthias will send a follow-up to non-responders.

Item #4 – New Client Communication Policy Draft

- See attached.

Item #5 – Next Meeting

- December 15, 2019, 10 a.m. in Orlov Board Room

4001 LANCASTER DR NE, SALEM, OR 97309
T 541-555-5555 U WWW.HLO.COM

Formatting

Rules and technicalities often determine the format of a meeting. If you are in a meeting and hear phrases like "I move to table this topic" or "I move to adjourn," you are taking part in an eighteenth-century tradition defined in *Robert's Rules of Order*. This book defines parliamentary procedures, a set of practices designed to get a group of people to debate a topic and arrive at an agreement. You are most likely to encounter this meeting format in government or corporate board meetings.

Meetings follow this general format:

- » **Call to order:** The person holding the meeting gets everyone's attention to begin.
- » **Roll call:** The person taking the minutes records who is in the meeting.
- » **Approval of minutes:** If this is a recurring meeting, the notes from the previous meeting are reviewed, approved, or corrected.
- » **Agenda items:** The meeting proceeds through the topics listed in the agenda.
- » **Adjournment:** The person running the meeting announces the official end of the meeting.

Business Letters

Business letters are formal documents that are a significant part of the professional world. They are the letters you use to gain employment, build professional relationships, and create opportunities. That's a ton of pressure to place on a single document, but the fundamentals of technical communication can guide you in creating a message and purpose that clearly speak to your audience.

Professional Use

Business letters require an appropriate format and tone for an audience that is, more often than not, external. Figure 7.6 shows the basic arrangement of elements in a business letter. As shown in this model, business letters are often printed on company letterhead. The paper is usually higher quality with features meant to impress the audience and represent the company's brand.

Figure 7.6. Business Letter Template. Business letters are generally created for an external audience. Because of this, they tend to be the most formal of all workplace documents.

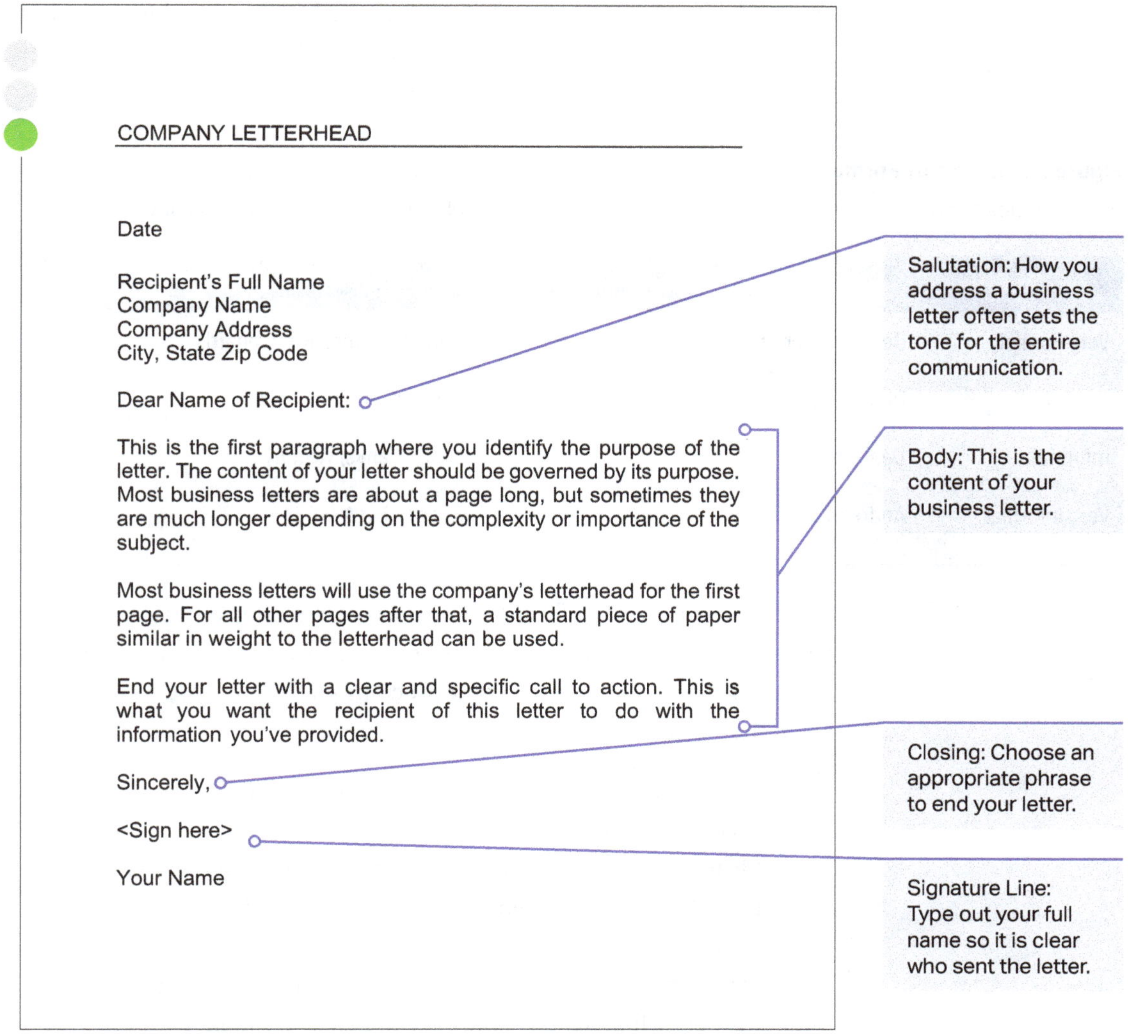

COMPANY LETTERHEAD

Date

Recipient's Full Name
Company Name
Company Address
City, State Zip Code

Dear Name of Recipient:

This is the first paragraph where you identify the purpose of the letter. The content of your letter should be governed by its purpose. Most business letters are about a page long, but sometimes they are much longer depending on the complexity or importance of the subject.

Most business letters will use the company's letterhead for the first page. For all other pages after that, a standard piece of paper similar in weight to the letterhead can be used.

End your letter with a clear and specific call to action. This is what you want the recipient of this letter to do with the information you've provided.

Sincerely,

<Sign here>

Your Name

Salutation: How you address a business letter often sets the tone for the entire communication.

Body: This is the content of your business letter.

Closing: Choose an appropriate phrase to end your letter.

Signature Line: Type out your full name so it is clear who sent the letter.

Formatting

As with other workplace formats, a business letter has a handful of simple expectations. You are probably already familiar with some of them. For example, most business letters use a standard opening (salutation) and closing (figure 7.7). The beginning of the letter signals the level of formality between the sender and the recipient.

Figure 7.7. Levels of Formality. Notice the levels of formality in these salutations and closings. The use of the term "dear" may seem like a term of endearment, but it is simply a convention used in business writing.

Formality	Salutations	Closings
Very Formal	Dear Senator Andrea Jensen-Joli:	Cordially yours, Kind regards
Formal	Dear Ms. Jensen-Joli:	Respectfully, Sincerely
Informal	Dear Andrea,	Best, Thanks, Warmly
Very Informal	Andrea,	Cheers, Later

There isn't one "right" way to format a business letter. Often, an employer has a preferred format, so ask for an example or use a business letter you received as a model. Here are the most common formats you will encounter in the workplace (figure 7.8):

» **Block format**: All text aligns with the left-hand margin with a space between paragraphs.
» **Modified block format:** The opening address and the closing are placed along letter's center line to distinguish them from the body of the letter.
» **Indented format:** Instead of an extra space between paragraphs, this letter indents the start of each new paragraph.

Figure 7.8. Business Letter Format. Consistency and alignment are important to consider when formatting your business letter. Pick one format and stick with it.

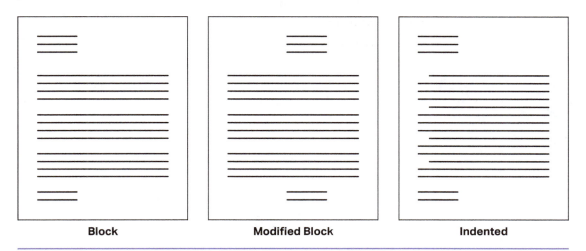

Block   **Modified Block**   **Indented**

Text Messages

We all know how to text, some of us better than others. Text messages are a common type of personal communication that is becoming more common in workplaces as well. You might even work for a company that pays for your cell phone and expects you to be available 24-7.

A text has more in common with speech than written language, according to Dr. Caroline Tagg, a British linguist who began studying text messages in 2009.[2] We use informal words such as "dunno," quick phonetic spellings, and ideograms (emoji) to create a personal tone we'd normally use when speaking in person. However, if you're sending texts as part of your job, you may want to take a more formal approach.

Texting is great for quick back-and-forth messages. It combines the benefits of spoken and written language into a compact message that can be sent and responded to almost instantaneously. Unlike a face-to-face conversation or phone call, texting creates a record, and you can put the conversation on pause while completing a task.

Professional Use

Because you text every day, it can be easy to forget that you can't text professionally the way you do with your friends. Be aware of the communication style used in your professional environment. Each workplace will have expectations for communication styles, tone, and level of formality. If you fail to recognize the need to adjust your communication style, even if the change is slight, the result could be confusion, as these sample text messages demonstrate.

Figure 7.9a. Ineffective Text Message. The first text message makes assumptions that the recipient will immediately recognize who is sending it.

Figure 7.9b. Ineffective Text Message. The second text message is one long stream of thought that is littered with typos and does not consider readability for the recipient.

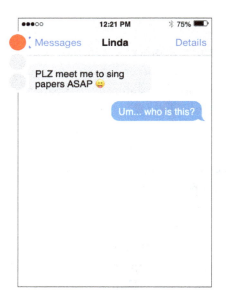

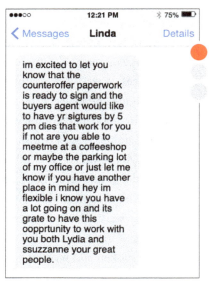

What's happening here? This is a realtor (Linda) asking her clients to sign a counteroffer on the home they're selling. Neither of these texts represent a career-ending mistake, but they do chip away at her credibility. Linda's clients trust her to handle complex transactions that involve their finances and determine how soon they can make the move into their new home. When a message strikes the wrong tone or creates confusion, it can slowly erode that trust.

Let's give Linda another shot. Here's a professional text that strikes the right tone (figure 7.10). The message isn't exactly using the kind of language you'd want to use in a business letter, but it's perfect for a text message, assuming she's not sending it after her clients have gone to bed.

Formatting

Text messages are designed to be brief. The technology that allows messages to be sent by cell phone is called short message service (SMS), which can contain up to 160 characters. You can, of course, text more, but the message is often broken up into multiple smaller messages. When texting professionally, limit yourself to short messages.

Figure 7.10. Effective Text Message. This text message is casual yet provides information so the recipient knows how to respond.

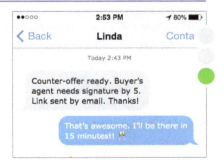

Communicating Professionally

How you choose to communicate in the workplace influences how you are perceived by your coworkers, bosses, and customers. You want to make a positive impression, so think carefully about the key factors in this section. These factors include using the appropriate tool, using the appropriate tone, evaluating content, prioritizing information, considering audience, and being persuasive.

Use the Appropriate Tool

You need to communicate with a coworker, but which tool do you choose? Should you send an email, compose a text, pick up the phone? Those methods have been around for a while. But today's workplaces have even more options as a result of web-based communication apps. Tools like Basecamp, Slack, Zoom, and Trello are common communication management systems you might encounter. Some businesses have their own communication platform created for that specific workplace. Given all of these different methods of communicating, how do you pick?

The first thing to remember about choosing a communication tool is that many workplaces have specific rules or guidelines about this. If you haven't reviewed the employee handbook, you should consult it. Beyond a written policy, most businesses have unspoken expectations about communication. If you're new to a place of work, check with a coworker or even a manager to determine the preferred communication methods.

The purpose of these apps is to improve the efficiency of workplace communication with a dedicated channel that isn't cluttered by distractions and outside requests. But sometimes a tool designed to create focus can cause distraction. Learn how to operate the apps efficiently, including when and how to turn off the alerts that interrupt you every five seconds. Like all other forms of workplace communication, these tools require you to understand your audience's expectations, which in this case are the expectations of your coworkers and business partners.

Use the Appropriate Tone

You know this by now, but we're going to remind you again: do not communicate in a professional setting the same way you communicate with your friends. You need to sound friendly without being overly familiar. But you also need to sound like a human, not a robot. How do you strike the right balance?

Tone is communicated by the words you choose and by your attitude toward the topic or audience. If you were ever told as a child to "watch your tone," you know it means to adjust your attitude. The same happens in writing. It's easy to slip up and use a tone that doesn't suit the message, whether from exhaustion, indifference, or frustration. Allowing these feelings to seep into workplace communication will eventually cause problems. Take a look at the different ways tone becomes clear in these examples (figure 7.11).

See Chapter 10 for more on tone.

Figure 7.11. Examples of Tone. Small changes in word choice can affect the tone of your communication.

Problematic Tone	Improved Tone
I just think that maybe this project is moving in the wrong direction. *(Uncertain)*	This project is moving in the wrong direction. *(Confident)*
It's imperative that you get the form back to me ASAP. And don't forget to sign it! *(Rude)*	Please sign the form and return it to me by Friday. *(Respectful)*
Your paperwork is a mess. *(Biased)*	Your paperwork is incomplete. *(Objective)*
Your recommendation has been received. You will be contacted about the next steps soon. *(Roundabout)*	I received your recommendation and will contact you next week to discuss the next steps. *(Direct)*
I wish you would have told me sooner. *(Negative)*	Thank you for telling me this now. *(Positive)*

Exclamation points are another way that tone can be expressed in writing. This flamboyant punctuation mark adds punch to emotion-filled statements, so it's rare that you would need it in more formal workplace communication. However, less formal modes of communication might benefit from the occasional exclamation.

For example, you text your boss to let her know you've got the flu and won't be able to make it in. You receive her response: "Oh no." Without an exclamation point, her message might come across as flat, almost deadpan. The key is to make conscious choices in your writing to create a tone that helps you get your message across and reduce confusion.

Evaluate Content

Your workplace communication tools are not for personal conversations. Most people would find it acceptable to email your colleagues about a get-together after the end of a hard project. However, use your best judgment when sending a message that will be out there forever. Ask yourself two questions before you hit send: Is this work-related? If not, should I be using my personal device instead?

In 2018, the U.S. witnessed FBI agent Peter Strzok get questioned by members of Congress. Why? He sent texts to a friend (and colleague) making fun of the president. The problem—at least one among many—is that he was using his work system to make some of the jokes.[3] If you want to stay away from a congressional committee, we recommend you keep the jokes off your work email.

A sense of humor is a good, healthy thing. However, jokes do not always translate well—especially if they require a solid grasp of tone. If you send a joke via text or email, remember that the receiver can't see your face or hear your voice. Does the joke work without you there? If the answer is no, maybe don't send it. Would you tell the joke in front of some children? No? Maybe rethink your options. We'd like you to keep your job.

Prioritize Information

If you understand your purpose and know your audience, then prioritizing your document's content is easier. Before you give a speech, send an email, or ask for a raise, identify your top two or three main ideas. For example, if you create an email announcing an upcoming event at your workplace, but you fail to mention the date, time, or location of the event, you have a problem.

Lead with the information your audience needs most. In this case, your email should not only mention the date, time, or location of the event, but it should make this critical information hard to miss. As often as possible, open with the most important information in a workplace document and follow with any other useful details.

When you're working fast and have multiple projects you're responsible for, you might find it challenging to identify your priorities. That's why it's best to think strategically when you compose a memo, email, business letter, or any other professional communication. In general, keep your writing focused on a single point, organize your information clearly, and avoid unnecessary information.

Here are some guidelines to determine what's most important:

» Recognize the limits of your document and your audience's attention.
» Identify the key takeaway, the one thing your audience needs to know.
» List the information that supports this main idea.
» Organize this list by importance, sequence, or chronology.
» Keep it short, but provide a way (a website link or email address) for the audience to obtain more information if needed.

You *can* insert interesting side information in your documents or presentations—just make sure it's not interrupting or distracting from the main flow of information.

Consider the Audience

Take time to get to know your audience. Are there any cultural factors, such as expectations of formality, that you need to observe in order to make the message successful? What about addressing the person at the beginning of your message? For example, when someone earns the title "Doctor," they tend to prefer being addressed that way.

Other factors to consider are whether you are making assumptions about a person's identity based on their name, profession, or gender. Take a look at the signature field example, which is one way to glean useful information about a potential client or colleague (figure 7.12). What do you know about their identity? If J. didn't indicate any preference for pronouns, what mistake might you make?

Figure 7.12. Respectful Communication Practices. Using a person's correct pronoun is a way of showing respect. When in doubt, use the person's full name.

D. Stanley
Regional Representative
People with Anonymous Names (PAN), Inc.
d.stanley@paninc.org
(541) 555-5555
Pronouns: They/Them/Their

A significant part of avoiding mistakes when it comes to addressing your audience is considering your own biases. Do you treat someone differently because of their gender, race, ethnicity, age, physical appearance, or ability? Evaluating your own biases can be difficult, especially when it comes to implicit biases, those beliefs you've learned to act on without thought. To learn more about these unconscious attitudes, you can take an online test, such as the Implicit Association Test.

If you genuinely aren't sure about the appropriateness of a piece of communication, consider asking a trusted coworker for their opinion. Getting an outside opinion is helpful on many levels.

Collaboration

While you may have written most of your academic papers on your own, much of what happens in the workplace happens collaboratively. Students sometimes think collaboration and group work are the same thing. They are not.

In group work, a team of individuals is assigned a project, and each person works independently on their part of the project. Group members check in with each other, but they do their work solo. Often most of the work falls to one or two people, and these are usually the ones most invested in it. Students who have tried group work often complain that they could do better work completing an entire project on their own. Miscommunication, missed deadlines, and document inconsistency are common.

Document inconsistency is particularly problematic in the professional world. When the work of individuals is spliced together, the shifts in voice, style, and meaning can distract the user. Individual work often produces a better product because the various sections are more likely to be consistent.

However, collaborative work generally creates higher-quality products by blending the best characteristics of both group work and individual work. **Collaboration** is when a group collectively assigns tasks to each member, usually within their field of expertise, and every member must participate in order to complete a project. It means that the group shares a common purpose and message that appeals to an identified audience. Collaboration includes consistent check-ins and allows the participants to access higher levels of thought and creativity than they could on their own.

Figure 7.13 illustrates the difference between group work and collaboration. Notice the visual distinction between the documents. Group work tends to produce a document that is characterized by distinct sections that do not form a cohesive unit. Collaboration, on the other hand, tends to produce a blended document that forms a unified whole.

Communication is the foundation of collaboration. When communication is going well, problems become opportunities for creative exploration. This means respecting the other participants' views and ideas. Because technical communications is often a global endeavor, collaboration requires that you respect cultural differences.

Figure 7.13. Group Work vs. Collaboration. Group work and collaboration are sometimes used interchangeably, but they do not mean the same thing. As a professional, you need to seek out productive partnerships with your colleagues.

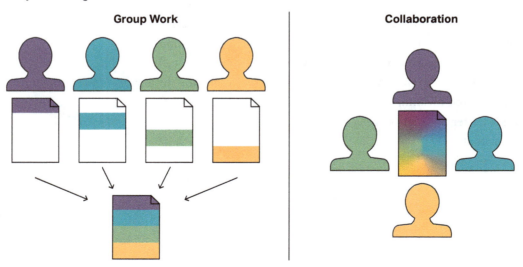

Group Work   **Collaboration**

Communication

Breakthroughs often happen in the spaces where collaborators see things differently. Ideally, you should view conflict as an opportunity for growth. When a team allows its members to engage in constructive and respectful disagreements, the result is collective ownership of what is created.

Varying commitment levels can tip the dynamics of the process in ways that might send you hurtling back towards group work. To keep your collaboration on track, talk with your team members about what success looks like for your project. Time frames and deadlines form an important part of collaboration. Participants need to communicate regularly about if and how they will meet their deadlines.

Differences of opinion can be difficult to navigate in a collaboration, but they are also important. One of the reasons group work often fails is because it typically avoids controversy, resulting in a document that is inconsistent in its message. Resolving differences through collaboration takes hard work and effective communication, but the result is typically a stronger document. If you run into conflicts within your group, try the following strategies for conflict resolution:

» Characterize the issue in clear terms.
» Agree on a common goal.
» Establish ways to meet the shared objective.
» Determine what stands in the way of resolution.
» Acknowledge the best solutions.
» Agree on each team member's responsibilities.

Methods for Collaboration

Many types of collaborative software exist, and your employer will most likely have a preference for which one your team uses. Most collaborative software programs share common tools, such as one-on-one or group options; real-time collaboration and document sharing; planning, tracking, and reporting tools; and shared permissions. See figure 7.14 for additional strategies for collaboration.

Figure 7.14. Collaboration Strategies. This chart shows the strengths and weaknesses for different approaches to collaboration. Often a team cycles through multiple approaches during a single project.

Approach	Definition	Best suited for	Strengths	Weaknesses
Divided	The team splits up the task and assigns responsibilities for each part to individual members	Generating drafts	» Produces a high volume of content quickly » Reduces face-to-face time constraints	» Can result in an uneven style and inconsistent content
Collective	The team sits down together at the same time to draft, ideally in person	Contained writing, such as outlines	» Produces a unified style and solid content » Results in a high level of ownership	» Can be time consuming, especially when team members disagree on approach and content
Solo	One person assembles the notes from the team and creates the final document alone	The workplace where the individual has been hired for this specific purpose	» Produces a strong, unified document if the solo writer is skilled » Reduces the team's time commitment during the drafting process	» Leaves the burden of the project on a single person » Can result in a poor document if the individual is not sufficiently skilled
Role	Team members take on different roles based on their strengths	Teams where each person is bringing a specific expertise	» Efficient » Results in a strong document with a high level of ownership	» Can devolve into group work » Can place the burden of the project entirely on the writer, similar to the Solo approach
Freeform	Team members share, add to, and edit a document that is available online (i.e., Google Docs)	Teams who cannot meet in person	» Allows for distance work » Flexible	» Progress can stall when team members disagree on approach and content » Challenges with quality control

Global Workplace Etiquette

Many technical communicators create documents and deliverables for global audiences. In order to be competitive in the workplace, you must prepare for different styles and approaches to business etiquette.

Here's a brief list of business etiquette to get you started:

» **Arrival time:** Punctuality is a priority in the U.S. and Germany, but other countries may have a more casual relationship with time. To be safe, show up on time, but be patient if you're kept waiting.

» **Names and honorifics:** Naming conventions differ around the world. For example, workers in the U.S. generally prefer using first names, a more casual address. Many other countries, however, favor the formal equivalent of "Mr." or "Ms." accompanied by a surname. Don't make assumptions about gender or marital status. Ask people what they prefer to be called.

» **Greetings:** A firm handshake is appropriate in the U.S., but not in France. In Europe and the Americas, it's standard to present a business card with one hand. But people from the Far East, Southeast Asia, or the Indian subcontinent tend to offer business cards with two hands extended. Watch and mirror the exchanges of others if you're not sure.

» **Meeting protocol:** In some parts of the world, people like to get business transactions out of the way before socializing. Other regions value getting to know a person before conducting business. Research can tell you whether chitchat, interruptions, or following an agenda are customary.

» **Correspondence:** It's worth taking your time when you're communicating with a global audience. Nothing says lack of professionalism more than a hasty email. The preferred communication style might be direct or indirect, casual or formal, concise or chatty, but it takes time to figure this out. Also, be mindful of time zones and when the recipient will likely receive your message.

Do some research on your own before you travel. Research countries and their culture so you don't embarrass yourself or the company you represent. Each country is unique, and it shows respect if you are informed.

 ## Case Study

Meeting Etiquette

This case study is an opportunity for you to put into practice what you've learned. Part of this chapter focuses on using technology to communicate effectively at work. Look at the following case study to consider how the concepts of technical communication, specifically audience, message, and purpose, apply to a virtual meeting:

Let's revisit Frank, the executive assistant, from earlier in the chapter. Frank is in charge of setting up virtual meetings and keeping minutes for his manager. It's his responsibility to make sure meeting participants have everything they need. Frank is meticulous about the number of chairs around the table. He checks the temperature of the room and makes sure everyone has water. He even double checks the phone and camera power settings. Once all of the participants have arrived, Frank announces that he will keep the meeting minutes. He feels confident that he has thought of everything.

Frank's manager initiates the video call to their international offices. Frank notices that the camera is focused on the center of the table, which means his boss is not visible to the video attendees. He also notices that new employees in the international offices have visible name plates in front of them. He gets a sinking feeling he could have done more to make the meeting go smoother.

Discussion

» While Frank made sure his colleagues in the room had everything they needed, what could he have done for the virtual participants?

» What protocols do other countries have for meetings? Do a quick internet search on business etiquette in other countries.

» What five items should be included in Frank's meeting minutes document?

Checklist for Workplace Communication

Internal Communication

☐ Do you make documents and knowledge easily accessible?
☐ Do you understand and use language consistent with the company culture?
☐ Have you identified the company's goals?
☐ Do you offer recognition for a job well done?

External Communication

☐ Do all documents share recognizable branding?
☐ Do documents use consistent layout and design?
☐ Are documents proofread before they're sent out?

Personal Communications

☐ Do you communicate ideas responsibly and are you open to constructive criticism?
☐ Do you use an appropriate tone in your communications?
☐ Are you aware of your body language in all interactions?

Conclusion

When it comes to workplace communication, a lot is at stake. Your ability to communicate with your colleagues or clients significantly impacts your success. While a lack of skill in workplace communication could be detrimental to your reputation, excelling in this area can distinguish you as a valued employee. Fortunately, you can gain confidence in workplace communication if you focus on meeting the expectations of your work environment and following established communication formats.

This chapter gives you a start by pointing out common expectations. You'll need to continue to evaluate the expectations for your specific workplace. In addition, you'll want to stay informed about the current state of preferred workplace tools as they change over time. By focusing on the fundamentals of technical communication—audience, message, and purpose—you can be ready to adapt to whatever new form of communication comes your way.

Notes

1. Jack Birkus, "Internet Access at New High Wordwide Before Pandemic," *Gallup*, April 8, 2020, https://news.gallup.com/poll/307784/internet-access-new-high-worldwide-pandemic.aspx.
2. "Texting Is Closer to Speech than the Written Word, Claims Academic," *The Telegraph*, August 7, 2009, https://www.telegraph.co.uk/news/science/science-news/5984225/Texting-is-closer-to -speech-than-the-written-word-claims-academic.html.
3. Adam Goldman and Michael S. Schmidt, "FBI Agent Peter Strzok, Who Criticized Trump in Texts, Is Fired," *The New York Times*, August 13, 2018, https://www.nytimes.com/2018/08/13/us /politics/peter-strzok-fired-fbi.html.

Chapter 8

Technical Definitions and Descriptions

Abstract: Technical definitions and descriptions give meaning to a process or procedure and can be found in a variety of technical documents. Technical communicators generally use three types of definitions: parenthetical, sentence, and extended definitions. A description is a kind of long-form definition that requires precision. In order to create a useable definition or description, you must know who is going to use it, when they're likely to use it, what they need to know, and what they already know. Audience plays a big role in determining the amount of information you need to create effective definitions and descriptions.

Looking Ahead

1. Why Definitions and Descriptions Matter

2. Creating Definitions and Descriptions

3. Parenthetical Definitions

4. Sentence Definitions

5. Extended Definitions

6. Descriptions

7. The Known-New Contract

8. Legal and Ethical Implications

Key Terms

» Chain Method
» characteristic
» class
» concrete language
» context
» definition
» description

» extended definition
» Fork Method
» glossary
» jargon
» Known-New Contract
» name
» negation

» parenthetical definition
» process description
» product description
» sentence definition
» user profile

Why Definitions and Descriptions Matter

When you were in grade school and didn't know the meaning of a word, your teacher most likely told you to look it up. Now, as a technical communicator, it's your job to know the terminology of your specific field, to use it accurately, and to be able to explain it precisely to someone else. You can't tell your audience to grab a dictionary. You *are* the dictionary.

The good news is that you don't have to know everything. But you do need to know how to use your expertise to define unfamiliar terms and situations. A **definition** is a short explanation of the meaning of a word or phrase. Technical definitions help users understand an unfamiliar term or clarify a word that has multiple meanings. A **description** is an extended explanation that focuses on physical characteristics of an object or situation. For an object, a description might include its size, weight, shape, color, material, and use. For a situation, the description might include its purpose, frequency, duration, location, and other factors.

Though the specific words you need to create a definition or description will depend on your audience and your industry, the way you create these technical elements will follow a standard format, which we'll cover in this chapter.

As an expert (or expert-in-training), you need to determine how and when to provide definitions and descriptions to your audience. Never assume that a user knows everything you know. Keep in mind that your background and education have given you a specialized vocabulary. When you understand a subject thoroughly, it can be difficult to imagine what someone new to the subject knows or doesn't know. To combat this tendency, keep these questions in mind:

» What does my audience need to know?
» How much detail does my audience need?
» How will my audience use this information?

In your own projects, your job is to determine if the user needs a quick reminder of a term's meaning or an extensive description of an object's parts, purpose, and function. To illustrate this, we'll revisit Leticia and

Jason's board game project, *Alpacas of Doom!,* from earlier in the book. Their game uses a variety of familiar concepts (dice, alpacas) with adjustments for gaming purposes. To communicate these new ideas to their boss and potential funders, they need to define how and when these modified elements are used. Documents that alter familiar items and introduce new concepts present an ideal opportunity for exploring definitions.

Creating Definitions and Descriptions

Effective technical communication includes the ability to define terms, describe methods, and clarify concepts. With specific and understandable definitions and descriptions, you can make a document more useable.

In Chapter 1, we introduced four guidelines to help your technical documents become more useable:

» Be clear
» Be precise
» Be concise
» Be accurate

Think of these concepts as four points on a compass. Taken together, they will orient your users and get them where you want them to go. As you create content, it's wise to look up every so often and orient yourself, too.

Definitions and description often go hand in hand. Clear definitions require some description to be effective, though they may also rely on explanation and context. **Context** is the setting in which the word appears. For example, in a technical document, the word "screw" likely refers to a small, spiral-shaped metal piece used to attach two or more solid pieces together. In a text message from your angry friend ("screw you!"), the word is used in a different context and has a completely different meaning.

If the user has the potential to interpret words in too many ways, your document won't succeed. In the worst-case scenario, a failed communication can have severe consequences.

Be Precise

See Chapter ❶ for more about precision.

A precise definition is one that requires exact and specific language, which can be more challenging than it seems. Beware of revolving door definitions such as this one: "Precision means to be precise." Using a term or a variation of a term to define the term itself sends your user in a confusing circle. Your goal should be to explain the new term using words that the user is already familiar with. So what precisely does "precision" mean?

Figure 8.1 defines "precision" using different definition strategies. In some cases, a term or concept may require an extended, multi-paragraph definition that includes deep research. For instance, we took a little more time to define "precision" here because it's a word that many people think they already know. In technical fields, precision is essential and has a specific meaning that's different from its common usage. We break the term down for you in this example to demonstrate its importance. Additionally, we use this term to show how you can begin making your definitions more precise.

For example, *Alpacas of Doom!* players use 3d6 to determine game effects. This language only feels precise to people who already know what 3d6 means. A more precise description would recognize that most board game enthusiasts know that 3d6 refers to a set of three six-sided dice, but newer players might need the additional description. This is why games that have been around for decades still include glossaries and descriptions—it's no fun if you don't know the rules of the game.

Figure 1.1. Definition Strategies. To define a term, break it down into its simplest parts. What is it made of? How is it used? What is it like? You can offer further explanation of a term by exploring what it is not or by using examples, and analogies.

Comparison/ Negation	Precision is not the same thing as accuracy. While accuracy measures how close something is to being true or meeting a standard, precision measures the degree of exactness with which an operation is performed.
Examples	A precision instrument is a tool that can be controlled with exactness to produce a specific result.
Analogy	Precision is the ability to hit the bull's-eye on a target again and again.

Target Your Audience

How can you be sure you're hitting the mark? In technical documents, your audience is the bull's-eye, the center of your communication target. Always make sure your definitions answer the questions "What is it?" or "What does it involve?" in ways that make sense for your audience. As mentioned earlier, you might need to include analogies, examples, or comparisons to clarify your meaning.

One strategy for understanding your audience better is the creation of a **user profile**, a document that collects information about your audience from interviews, surveys, reports, or conversations with your client. The more you get to know about your audience, the easier it will be to decide what level of detail to use, what you need to define, and what can be left unsaid.

When your technical document has more than one audience segment, definitions can help you bridge the gap (figure 8.2). For example, in an audience assessment report for investors, Jason and Leticia identify that *Alpacas of Doom!* will most likely appeal to gaming enthusiasts who enjoy themed resource collection board games. They are the same crowd that is likely to already own games such as *Settlers of Catan* or *Zombicide*. To guarantee their document's usefulness, Jason and Leticia need to take this audience's experience into account and also recognize that some members of their audience may be entirely new to gaming.

Figure 1.2. Technical Knowledge in Audience Segments. Technical communicators often translate technical information from one audience segment to another.

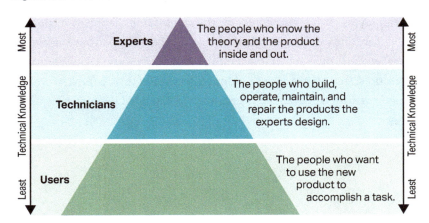

Recognize Jargon

Industry-specific terminology, or **jargon**, requires definition when communicating with an audience outside your field. You also need to define words that have multiple potential meanings or acronyms. For example, take the word "heap." To a computer scientist, it indicates a type of data structure, but to a layperson, it could mean a big pile. Be specific and determine what needs to be defined for clarity.

You should develop the habit of providing your own definitions instead of relying on dictionary definitions, which may not be specific enough for your communication needs. For example, if you're writing a technical report on nanotechnology, you'll find Oxford's *Dictionary of Mechanical Engineering* more useful than Dictionary.com.

Jargon doesn't just apply to complex technical terms. For instance, "hood" has a variety of specific meanings. Is it part of a ventilation system? Does it cover the engine of a car? Is it the part of a jacket or sweatshirt you wear on your head? In a professional kitchen, the hood is the ventilation and fire suppression system placed above cooking surfaces that draws smoke out and away from the kitchen and dining area. In a document discussing the legal and safety requirements for opening a restaurant, users need a specific definition to make sure they understand this industry-specific term.

When to Provide a Definition

The best time for a definition is immediately before or after the first use of the term. Whether or not you need to provide a definition depends on your audience. When you communicate with other experts in your field or employees in the same company, you can assume they share a common vocabulary and can leave out definitions to save time. If you have any reason to believe someone may not be familiar with a term, you should provide a definition. Three types of definitions we'll cover in this chapter, in order of complexity, are parenthetical, sentence, and extended definitions.

Parenthetical Definitions

When someone needs just enough added information to understand a term or acronym, a parenthetical definition is the best choice. A **parenthetical definition** provides an explanation of a term immediately after its first use, typically enclosed in commas or parentheses. For example, documents that deal with health concerns for a large audience may refer to "seasonal influenza, commonly known as the flu," or simply "seasonal influenza (flu)."

Parenthetical definitions often provide a synonym (a word or phrase with a similar meaning) enclosed in parentheses or commas immediately after an unfamiliar term. The previous sentence shows this definition strategy. This approach allows the specific and general meaning of a word or phrase to coexist within a single sentence and reach a wider audience.

One way to think about parenthetical definitions is to see them as handrails that guide the user. Your goal is to get your user from point A to point B, from the unfamiliar to the known. Parenthetical definitions can help you get there without having to explain too much and risk getting off topic. Parenthetical definitions are intended to interrupt the flow of a sentence as little as possible. Users familiar with the term can skip over them. But users who need a little extra assistance will appreciate that the definitions are there for them.

When to Use

A common use for a parenthetical definition is to introduce an acronym for the first time. The usual practice is to spell out the term in full and follow immediately with the acronym. Afterward, users will understand the acronym when they see it. For example, if you use meta-analysis from the National Institute of Mental Health (NIMH), you can introduce the institute followed by the acronym as you see in this sentence. In subsequent references, you can use NIMH instead and avoid taking up valuable space in your document.

Textbooks often use parenthetical definitions to make sure the user has quick and immediate access to important vocabulary. Acronyms (abbreviations) or specialized terms are defined parenthetically (in passing) when the audience includes people who might not be familiar with these terms. Parenthetical definitions are also used to make sure the audience is guided toward the correct understanding of a term in order to avoid misinterpretation.

How to Use

Parenthetical definitions are best used where the audience needs just a little more clarification or when space constraints make longer definitions impractical. Keeping a parenthetical definition to five words or fewer is a good guideline. This isn't a hard and fast rule, meaning it isn't an absolute requirement. However, if you need more than five words, you may want to use a sentence definition instead.

Sentence Definitions

A **sentence definition** provides an explanation of a term embedded in a sentence. In fact, the sentence you just read provides an example for how to structure a sentence definition. Sometimes the term is *italicized* or **bolded** to indicate to users that a definition can be found within the sentence.

A sentence definition is ideal for the user who needs a little more explanation than a parenthetical definition provides but who doesn't need the deep dive provided by an extended definition. The use of bold or italicized text allows a user to scan a document for unfamiliar terms. Textbooks like this one use this approach to make it easy for their users to find the information they need to remember.

When to Use

Sentence definitions are useful in a **glossary**, a list of defined terms usually located at the end of a document. Sentence definitions are also useful any time someone needs a moderate amount of explanation. They can work as a topic sentence for a paragraph by defining the paragraph's main idea followed by supporting sentences that give more in-depth information about the topic.

See the glossary at the end of this book.

How to Use

Sentence definitions often use a pattern that involves identifying the name, class, and characteristics of your term. These three elements form the foundation of a sentence definition (figure 8.3). **Name** refers to the specific term, thing, or concept you are defining. For example, "alpaca" describes a specific animal name, but what makes it different from, say, a llama? To answer this, you need to move on to the next category.

Class is the more specific category for the word. It does not simply name the object but provides additional details to help us classify it. For example, an alpaca is a mammal that belongs to the camel family. Llamas, camels, and alpacas are all part of the same family, though they are clearly different at the species level.

Figure 1.3. Elements of Sentence Definitions. Using the name, class, and characteristics of the unfamiliar term, each category becomes progressively more specific as it expands on the category that came before it.

Name	Alpaca
Class	Mammal in the camel family
Characteristics	Indigenous to South America, smaller than a llama with rounder hips

Characteristics are the unique traits that make the term or thing stand out from other terms in its category.

Creating a table, like the one in figure 8.4, can help you craft a clear definition for your audience. You can use the table to map out the sentence definitions you need in your project. After you've mapped out the definitions, you are ready to craft them into complete sentences.

After Jason and Leticia build their definition table, they write the appropriate terms into sentence definitions: "Alpacas are small, determined, civilized explorers who venture into the wasteland to find valuable lost technologies that will help their families and friends."

A glossary, like the one shown in figure 8.5, is written in the style of a dictionary entry. If you are working on a project with multiple people or creating a document that will be translated into another language, a glossary can help you maintain consistency of style, voice, and use of terms. For instance, the sample glossary in this section helps Leticia and Jason use the same language when they describe characters in their game.

Figure 1.4. Sentence Definition Table. In this table, game designers Leticia and Jason generate definitions for terms used in their game. The categories of name, class, and characteristics assist in the creation of definitions.

Name	Class	Characteristics
Alpaca	Wasteland denizen	Civilized, explorers, small but determined, seeking tech/treasure, social
Eagle	Unmanned aerial vehicle	Used for surveillance missions
Blaster	High-tech weapon	Leaks radiation when used, but delivers high damage
D6	Type of dice	Standard, six-sided

Figure 1.5. Sample Glossary. Notice that a glossary definition does not require you to write in complete sentences. The definitions should follow the same pattern and use parallel language to introduce the terms.

Alpaca: clever, techno-savvy, Andean relative of the llama that enjoys going on adventures in search of lost technologies from ancient civilizations
Blaster: powerful radioactive energy weapon
D6: six-sided die
Eagle: unmanned aerial vehicle used for reconnaissance missions
Quetzalcoatl: the feathered serpent from Mesoamerican myth and the final challenge in *Alpacas of Doom!*
Tripalpaca: all three dice in one roll showing the alpaca image

Check for parallelism by substituting the phrase "is a" in place of the colon.

Example: "Blaster is a powerful radioactive energy weapon."

Extended Definitions

An **extended definition** provides an explanation over multiple sentences and sometimes several pages. For example, this paragraph is an extended definition of the extended definition. Extended definitions provide the most context to the user and help solidify the document's purpose. This kind of definition goes beyond the sentence level by presenting clear facts, examples, and anecdotes the user will understand. It's important to meet users at their level of understanding before introducing new concepts.

When to Use

The need for extended definitions is determined by your audience's knowledge base. Extended definitions work best for situations that call for an in-depth approach. In general, users don't want to read an extended definition if they don't have to. Effective communication is the result of clear, concise, and purposeful writing.

A technical report is an appropriate place to make use of extended definitions. The purpose of a technical report is to inform an audience and make recommendations based on the findings. The audience of a technical report expects to see specific and precise information on the report's topic. This level of detail gives the audience confidence that they are getting a justified recommendation that will help them make an informed decision.

How to Use

An extended definition usually starts with a standard sentence definition followed by supplementary information and ends with a discussion of the term, phrase, or concept. If you go back to the process for developing sentence definitions (name, class, characteristics), you will notice that you can use the same process for building the first sentence for your extended definition. Here are two examples of extended definitions from students (figure 8.6).

Figure 1.6. Student Models of Sentence Definitions. Both student examples begin with a sentence definition. Example A loses its focus in the second sentence. Example B stays on track by answering these questions: What is it? How is it used?

This student model begins with a basic sentence definition.

Ⓐ

The TI-Nspire Touchpad calculator from Texas Instruments is a handheld algebraic graphing calculator that has programmable software capable of solving a wide range of equations. This calculator can perform many different types of equations such as factor and expand variables, compute limits and give exact solutions, and is solving graphic equation symbolically.

The next sentence expands on the types of equations, but the list gets confusing as it continues due to lack of parallelism.

Ⓑ

A foil cutter is a small, serrated knife attached to a wine key (a wine and bottle opener that resembles a Swiss Army knife). Its purpose is to provide a clean and safe cut while removing the foil at the lip of a wine bottle. In homes, a foil cutter removes the risk of potential harm when individuals try to use a knife (or some other kind of sharp object) to cut away the foil. In a business setting, foil cutters increase efficiency, increase professionalism, and reduce risk of potential harm to employees.

This student model begins with a strong sentence definition. It extends this definition by anticipating the user's questions.

The next step in writing an extended definition is to determine what in the first sentence needs further context and explanation. Often, examples of the concept can be useful. With these additions, you can develop the definition by discussing its position within the broader context. Consider including the origins of a word or phrase or background information. Sometimes it is useful to use **negation**, a description of what the term does *not* mean. At other times, a visual can help to extend your explanation.

Descriptions

Think of descriptions as an extreme version of the extended definition. They go beyond the initial questions of "What is it?" and "What does it entail?" to include other questions, such as:

» What are its parts?
» What does it do?
» How does it work?

Descriptions are used in a variety of documents to describe an object or process. In the workplace, you find them included in purchase orders and as parts of a longer technical document, such as manuals, reports, and instructions. A description focuses on specific details like background, features, physical attributes, function, qualities, and visuals.

Like definitions, descriptions use specific, accurate, and concise language to help the user better grasp the concepts in the document. Also, building a bank of descriptive phrases helps you and your collaborators provide accurate, consistent information.

When to Use

Descriptions are best used when you need to define a process or make the user aware of the parts they need. Often, visual aids like photos and diagrams accompany descriptions to help the user recognize and interact with the described item or process. Even without images to help, description is the tool we use to help someone else "see" the thing we're describing when they haven't encountered the item or concept before.

To understand the complexity of writing an effective technical description, imagine what it's like to explain 5G networks to someone who has never used a smartphone. To define this new concept, you may need to define other terms along the way. For example, you would need to explain that a 5G network is a form of broadband mobile communication that is wireless and allows the transfer of data from one mobile device to another at a rate faster than 4G. From there, you would need to define broadband, wireless, and 4G. Definitions require you to imagine what it is like to hear about something for the first time.

Here's another example from *Alpacas of Doom!* The game uses standard six-sided dice with images on four sides and two blank sides. This definition helps the user make a distinction between regular dice with pips (dots) and the regular dice used for this game. In the instructions, the next description would tell you how to use them and for what purpose.

How to Use

When crafting descriptions, use concrete language. **Concrete language** includes words that are sensory and tangible. Concrete language tends to be mostly nouns (things you can see and touch) and verbs (things you can do).

Use adjectives and adverbs sparingly in technical descriptions. Too many adjectives, especially in a continuous string, can cause confusion in surprising ways. It might seem like you're being specific when you use multiple adjectives, as in "5G broadband wireless mobile communication network." But this mouthful description will only make sense to people who already know what you're talking about.

One type of description you are likely to encounter is the product description. To develop a **product description** that details an object's physical characteristics, ask questions that focus on elements that you can see or touch. The object you are describing may require spatial and visual descriptions, temporal (time-based) and action descriptions, or any number of similar combinations. For example, if your job is to describe the features of a device, your description will require focus on size, shape, or color.

Examples of questions you might ask when writing a product description start simple and become more granular as you dive deeper into the product. For example, you might start with the question, "What is the first thing I notice about this product?" Chances are that what you notice first will also catch other people's attention first, which gives you a place to start when organizing your thoughts and drafting your description.

Other questions to ask yourself when writing a product description include the following:

» What are its distinguishing characteristics?
» What are the primary uses of this product?
» How is this product activated?
» Are there any clear dangers associated with using this product?
» What are the primary and secondary colors of this product?

In *Alpacas of Doom!*, Jason and Leticia want to use a simple dice system to add an element of chance. They asked themselves questions to figure out how they'd describe the dice to investors and manufacturers. This description is necessary for product development and will also help them understand game mechanics. They want to get the description right because it's not cheap to create and manufacture specialized dice (figure 8.7).

Another type of description you are likely to encounter is the process description. **Process descriptions** explain how complex events occur with attention to sequence, timing, movement, and necessary tools. Here are some questions to ask when developing a process description:

» What equipment is needed?
» Do users need to set up a space?
» What is the chronological order of each step in the process?
» What are the outcomes of each step?

Figure 1.7. Targeted Description. Jason and Leticia aim for precision in their description of the dice. The product description is for a specific audience: investors and manufacturers.

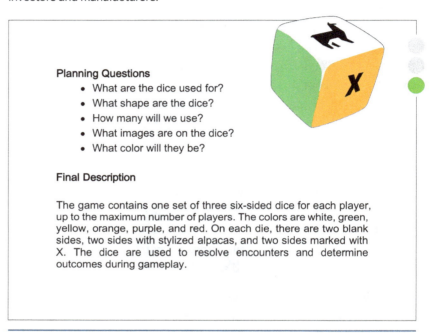

Planning Questions
- What are the dice used for?
- What shape are the dice?
- How many will we use?
- What images are on the dice?
- What color will they be?

Final Description

The game contains one set of three six-sided dice for each player, up to the maximum number of players. The colors are white, green, yellow, orange, purple, and red. On each die, there are two blank sides, two sides with stylized alpacas, and two sides marked with X. The dice are used to resolve encounters and determine outcomes during gameplay.

Since we just received Jason and Leticia's description of the dice their game uses, let's see a process in which these unusual dice are used (figure 8.8). Leticia and Jason answer questions to develop equipment for *Alpacas of Doom!* and consider further questions to determine how the equipment is used. These descriptions are important to players' understanding of gameplay as well as the overall theme of the game.

Figure 1.8. General Description. The goal of this process description is to make sure players can easily understand how to use the game's specialized dice. Simple, descriptive language helps players get to the fun part faster.

In *Alpacas of Doom!*, players use dice to resolve conflict. Players roll dice to determine the outcome of brawls with monsters, disputes with wasteland denizens, and missions for resources and technology. When one player fights a monster or haggles with a denizen, that player chooses another player to roll the dice to represent the opposition. Alpacas are civilized and engage in friendly competition, but they don't fight one another. Each die that rolls an alpaca is counted as one success. Blank results are "no effect" and X results cancel an opposing success. The player with the highest number of successes wins the encounter.	Essential information comes first.
	The description narrates the process.
	The description explains the end result.

The Known-New Contract

Effective definitions and descriptions benefit from using what the audience knows already and adding new information to that familiar foundation. Called the **Known-New Contract**, this basic principle of communication allows you to create a logical and easy-to-follow progression from sentence to sentence.[1] There are many ways to create definitions and descriptions using the Known-New Contract, but two of the most common methods include the Fork Method and the Chain Method.

In the **Fork Method**, each statement furthers the understanding of the topic. For example, in figure 8.9, the handle (A) is the topic: the connection between fast food and obesity. Each "prong" of this topic gives more information about the link between fast food and obesity. Prong B introduces more information related to the topic: "Studies have shown a strong association between fast food and obesity." Prong C adds more information to that: "These studies compare the number of fast-food restaurants and obesity rates per geographic area." Prong D adds a little more: "The research shows a significant increase in obesity in neighborhoods with a high concentration of fast-food restaurants."

Figure 1.9. The Fork Method. The Fork Method represents one way to make meaning using the Known-New Contract. In this method, the handle of the fork is the main topic (the Known), and the prongs provide additional information on the topic (the New).

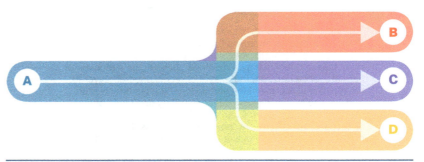

The **Chain Method** begins with a sentence with a topic accompanied by a comment on the topic. In this method, you build progressively on your audience's understanding. The comment at the end of your first sentence becomes the topic for the next sentence and so on, creating an interlocking chain of information. This method is one way to make your descriptions coherent and logical. Notice the variation in the phrasing from topic to comment in figure 8.10. These slight changes keep the explanation from becoming repetitive.

When to Use

Technical communicators frequently employ the Known-New Contract in all kinds of technical communication, but this principle is especially useful when writing definitions and descriptions.

When offering a definition, begin with what the user knows. By starting with a familiar concept, you create a pathway for the user to follow you into the more advanced concept. For example, when Steve Jobs introduced the iPad in 2010, he explained that it was "way better than a laptop, way better than a smartphone."[2] Jobs built on his audience's knowledge of the laptop and smartphone to sell his new idea.

Figure 1.10. The Chain Method. The Chain Method represents another way to make meaning using the Known-New Contract. Each link of the chain establishes new information (the New), which builds on information (the Known) previously introduced.

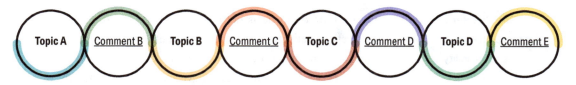

The Tilikum Bridge in Portland, Oregon (A), has crossing light art (B).

The lights (B) are controlled by the tide, speed, and temperature of the Willamette River (C).

The river's temperature (C) controls the color of the lights (D).

The hues (D) become orange and yellow if the river is warm (E).

How to Use

Begin with what you can confidently assume your user knows. This is your starting point. Don't begin with new, complex concepts, or you'll risk losing the user's attention. For example, Leticia and Jason will need to introduce the users of their game to original concepts. They should begin by reminding the user of familiar gaming concepts and then relate those familiar concepts to unfamiliar elements found in the game. A logical decision for Leticia and Jason would be to relate the concepts of the game to other games. This strategy works for both kinds of players: gaming enthusiasts and people who want a break from the family fights caused by Monopoly.

Legal and Ethical Implications

See Chapter 2 for more on ethical responsibility in technical communication.

Clear and accurate definitions help users navigate challenging technical issues. As a result, definitions carry legal and ethical implications for you as a technical communicator and for your employer.

The user of your document doesn't have access to your thought process. They can't see what you think, and they don't necessarily know what you know. Definitions help reduce the chance that users might misinterpret language in a document and make mistakes based on this misinterpretation. It's kind of like showing your work in your math homework.

Although there is such a thing as user error, it's a good idea as a technical communicator to assume that misunderstandings or confusion are not the user's fault. This assumption will help you stay sharp while drafting important documents that carry real consequences.

Conveying an accurate interpretation of facts is not only an ethical issue but also one of safety. If procedures and important terms are vague or incomplete and you work in a potentially dangerous environment, the consequences could be injury or death. That may seem scary right now, but it's a reminder to get a handle on writing clearly and accurately so that you are prepared if you ever need to take on a project with this level of responsibility.

Case Study

Alien Meets Alpaca

This case study is an opportunity for you to put into practice what you've learned. In any kind of writing, including technical communication, knowing your audience allows your message to be more effective. Look at the following case study to consider what needs to be defined and how much, what parts require a description, and how best can you explain all of this to the user:

During the early days of development for *Alpacas of Doom!*, Leticia and Jason challenged each other to describe elements of their game to a visitor from outer space. For example, how do you explain game rules to an extraterrestrial who doesn't know what a game is?

This exaggerated situation helped them get creative and think of new ways to describe something they had been working on for over two years. At the very least, it gave them something to laugh about after stressful meetings with investors.

Sometimes, in order to describe something effectively, you have to look at it with fresh eyes. Using the two types of description introduced in this chapter—process or product description—describe an everyday object to a visitor from another planet.

Remember, descriptions use specific, concise, and precise language. Your alien audience has traveled many light-years and perfected technologies well beyond human comprehension. In other words, do not talk down to your audience or you will be vaporized.

Consider the following items or processes to describe:

» Process: using an ATM, opening a door with a key, turning on a lamp
» Product: scissors, curtains, breath mints

Discussion

» First, define your purpose and audience so you can craft an effective message. How will that influence the narrative part of the description?
» How might the Known-New Contract help your process or product description?
» What potential problems might you encounter? How would you fix them?

Checklist for Definitions

Definition Basics

☐ Is the definition placed so it doesn't disrupt the flow of your content?
☐ Have you selected the best type of definition for your purpose and audience?
☐ Have you avoided using the word or phrase you're explaining in the definition?

Supplemental

☐ Have you tried describing your term by what it is not?
☐ Have you considered if well-labeled visuals can clarify the definition?
☐ Have you considered using hyperlinks in online documents to link to additional information?

Awareness

☐ Are your definitions grammatically correct?
☐ Have you considered the legal implications of your definitions?
☐ Have you considered the ethical implications of your definitions?

Conclusion

Throughout this chapter, we've used analogies to help you visualize the ways that definitions and descriptions work in technical communication. The parenthetical definition is like a handrail guiding users. The sentence definition is like a cornerstone that brings together two separate ideas and allows you to build upon them. The extended definition is like a bridge that takes the users to a place of understanding. The Known-New Contract underlies all of these concepts. Each sentence is a link in the chain that binds you, the technical communicator, to the user at the other end.

Notes

1. This principle was originally called the "Given-New Contract" in Herbert H. Clark and Susan E. Haviland, "Comprehension and the Given-New Contract," in *Discourse Production and Comprehension*, ed. Roy O. Freedle (Norwood, NJ: Ablex, 1977), 1–40.

2. "Apple iPad Launch: Live Coverage," *The Guardian* online, January 27, 2010, https://www.theguardian.com/technology/blog/2010/jan/27/apple-tablet-launch-live-coverage.

Chapter 9

Instructions and Procedures

Abstract: Some people mistakenly think that instructions and procedures are the same. Though these technical documents have similar parts, they have important differences, too. Instructions provide a series of detailed steps that define how to complete a task. Instructions exist so that the same actions can be repeated in the same order with the same result. Procedures, on the other hand, provide an overview of the best methods to accomplish a complex process. For both, technical communicators can benefit by focusing on how audience, message, and purpose interact. An essential step in creating effective instructions and procedures is usability testing. When you test a technical document, you can see how much the user understands, where you can make changes for greater usefulness, and how to avoid potential problems.

Looking Ahead

1. Why Instructions and Procedures Matter

2. Instructions

3. Procedures

4. Legal and Ethical Concerns

5. Usability Testing

Key Terms

- » alternate steps
- » anchoring
- » exploded diagram
- » fixed-order steps
- » instructions
- » nested steps
- » procedures
- » usability testing
- » variable-order steps

Why Instructions and Procedures Matter

Do you remember starting a new job? If you're like most people, when you showed up for your first day of work, you didn't know what to do. To whom do you report? What are you expected to know already? How do you record your hours? When is lunch? And most importantly, where is the bathroom? You didn't walk through the door knowing all this. Someone had to teach you.

It's normal to feel nervous about doing something unfamiliar, whether it's starting a new job or using a new piece of technology. This is where the technical communicator can help. Technical communicators create instructions and procedures to guide people through a range of actions, from individual tasks such as assembling a bicycle to complex, large-scale projects, such as the crew procedures for a rocket launch. Both document types require intense focus on the end users so they can carry out a task with confidence.

Instructions and Procedures Defined

Instructions explain how you do something. Typically, instructions are written for an individual user to complete a short task. Instructions use short, simple sentences with lots of action verbs that tell a person exactly what to do one step at a time. Often instructions feature detailed images.

Consider a piece of furniture that requires assembly (figure 9.1). You need instructions to know what tools to use, what parts are required, and what pieces to assemble in what order. Ideally, instructions take you from a pile of screws and boards to a finished nightstand with a minimal amount of frustration or smashed fingers.

Figure 9.1. Detailed Instructions. This set of instructions tells the user what to do and what not to do. Courtesy of Sauder.

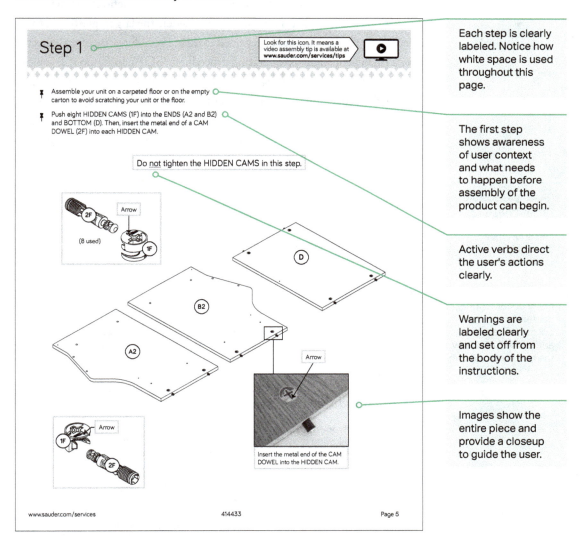

Each step is clearly labeled. Notice how white space is used throughout this page.

The first step shows awareness of user context and what needs to happen before assembly of the product can begin.

Active verbs direct the user's actions clearly.

Warnings are labeled clearly and set off from the body of the instructions.

Images show the entire piece and provide a closeup to guide the user.

Procedures differ from instructions because they focus on the bigger picture and define roles and responsibilities. While instructions deal with how to accomplish a specific task, a procedure considers the larger questions of *who* needs to do the task and *why* they need to do it. Procedures are often collected into a manual or standard operating procedures (SOP). In figure 9.2, you can see the first page of a multipage document that begins with a policy and then defines the procedures for carrying out that policy for a government agency.

Figure 9.2. Policy and Procedures. This model shows the first page of a multipage procedural document.

Headings separate the parts of the procedure and show their hierarchical relationship.

Many documents contain both policy (written rule) and procedures (how the rule is carried out).

LOCAL GOVERNMENT ABC

Accounting Policies and Procedures Manual
Policy #: PC-1
Last Revised: October 19, 2020
Policy Name: Petty Cash

1.0 Policy
Local Government ABC requires all departments to follow the below procedures establishing, overseeing, and closing out petty cash accounts.

Use of petty cash accounts is encouraged for purchasing of low-cost items from local vendors. Petty cash accounts will not exceed $1,000.

2.0 Procedures
The following procedures are designed to guide departments in establishing, overseeing, and closing out petty cash accounts.

2.1 Establishing a Petty Cash Account
2.1.1 Departments must complete the *Petty Cash Authorization* form and submit it to the Cash Management manager.
2.1.2 Once approved, department managers will be responsible for acquiring a lockable cash box. Contact the business services office for the box.
2.1.3 Funds and a transaction log will be provided once the lockable cash box has been obtained.
2.1.4 Each department will designate one employee to be responsible for the petty cash account.

You might encounter procedures in an employee manual in your work-place. Knowing how these documents function is beneficial. For example, many businesses have a procedure for reviewing employee performance. Such a procedure typically establishes how often an employee will be reviewed (annually, every six months, or every quarter), who will do the reviewing (the business owner or a department manager), the nature of the review (a test or an observation), the standard of assessment (points or pass/fail), and what will happen after the review (promotion, layoff, or continuing education).

In many ways, procedures help with decision-making, whereas instructions tell someone exactly what to do. While instructions are typically written for a single person, procedures are usually written for departments or even entire companies. Many procedures contain sets of instructions within them because an involved procedure often has multiple starting and stopping points. Take a look at this table that shows the main differences between instructions and procedures (figure 9.3).

Figure 9.3. Instructions and Procedures Compared. The differences between instructions and procedures become apparent in this chart when you consider the duration, audience, message, and purpose for each.

Instructions	vs.	Procedures
Describe shorter tasks that are completed in one session	Duration	Describe longer processes that often involve multiple tasks in succession
Describe only what the user needs to do	Message	Describe both what the users needs to do and why they need to do it
Focus on individual steps and actions	Purpose	Focus on an overall objective and decisions
Involve an individual	Audience	Involve more than one person

Instructions

Users turn to instructions when they need to complete a specific task. This might be something as simple as operating a coffee machine or something as complex as disassembling an engine. Regardless of the level of complexity, instructions need certain elements to be effective. Technical communicators must consider elements of effective organization, language, design, and visuals as they create instructions. As with all technical documents, instructions should be written to solve a specific problem for specific users.

Solving Problems

Instructions are clearly designed to solve a problem. Consider when a user is likely to encounter your instructions. Users tend to consult instructions only when they can't figure something out on their own or when something isn't working as expected. Talk about a tough audience. Your user is likely to be frustrated or impatient. They don't want to ask themselves, "Why am I doing this? Is this really the best approach?" Your job is to have already thought of those questions and to find the best answer.

Remember the situation and the needs of your user when you begin writing instructions. One of your challenges will be to remember that the audience hasn't thought about the topic as much as you have, particularly if the instructions are about an area in which you have significant experience. You may need to think back to the first time you attempted the task you are writing about. What did you need to know to complete the task? If you haven't completed the task, go try it and see what you learn. How would the activity be best explained? The ideal instruction set uses precise, brief language to save the user time and effort.

Don't forget to consider a range of circumstances that could impact how the user might interact with your document. Here are some questions to help you decide how to approach your document:

» How will users receive your instructions—in print, online, by mail, in person, cell phone, tablet, or computer?
» What types of visuals will enhance communication for users—photos, screenshots, clipart, drawings, diagrams, tables, charts, graphics, videos?

Best Practices for Instructions

The following section introduces several methods for making your instructions clear and easy to follow. Best practices include logical organization, standard and consistent language, clarity and accuracy, scannable design, and the effective use of visuals.

Organization

Every set of instructions describes a series of specific actions that must occur in a particular order. For example, "Insert bolt #1 into hole A." The user expects that following the instructions in the order presented will result in a successfully completed task.

To assist in this goal, make sure your instructions follow a logical sequence. Like a story, most projects have a beginning, middle, and end. Decide where to begin, and then break each part of the task into distinct steps by using headings. Describe the steps in order using short, simple directions. One way to prepare to write instructions is to use a template such as the one provided here to organize your steps (figure 9.4). This document works like an outline that you can expand to fit the needs of any instruction set.

Figure 9.4. Instructions Template. This template can be adapted for many kinds of instructions.

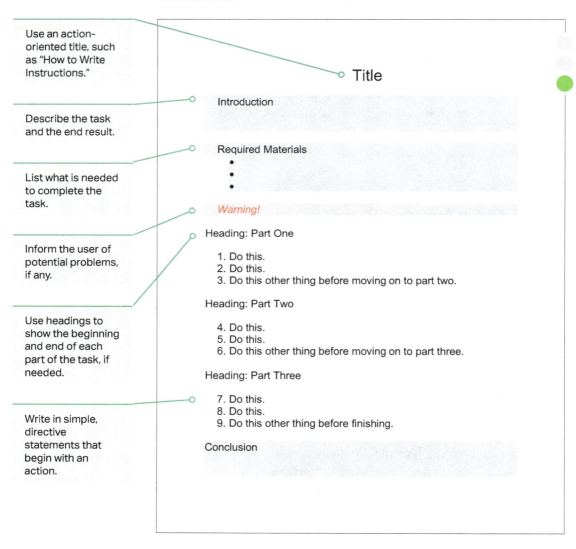

Use an action-oriented title, such as "How to Write Instructions."

Describe the task and the end result.

List what is needed to complete the task.

Inform the user of potential problems, if any.

Use headings to show the beginning and end of each part of the task, if needed.

Write in simple, directive statements that begin with an action.

Title

Introduction

Required Materials
-
-
-

Warning!

Heading: Part One

1. Do this.
2. Do this.
3. Do this other thing before moving on to part two.

Heading: Part Two

4. Do this.
5. Do this.
6. Do this other thing before moving on to part three.

Heading: Part Three

7. Do this.
8. Do this.
9. Do this other thing before finishing.

Conclusion

Standard and Consistent Language

Easy-to-follow instructions are a sign that the technical communicator has paid attention to word choice. In particular, the word choice should be standard and consistent. This means that the language of the instructions matches what the user is familiar with and expects. For example, you should avoid technical jargon in instructions meant for a general audience. If a specialized term is needed for accuracy, define the term right away with words or an image to help the user understand.

See Chapter **8** for more on definitions.

Consistency of language means keeping the same level of word choice and style throughout the document. You can ensure consistency by paying special attention to verb tense, verb mood, and parallelism:

» **Verb tense:** Keep your verbs in present tense—describe things in the now. For example, "Assemble your unit on a carpeted floor" and not, "It will be a good idea to assemble your unit on a carpeted floor."
» **Verb mood:** Use what is called the imperative mood—tell people what to do directly. For example, the sentences "Open the door," "stop the car," and "pass the salt" include verbs in the imperative mood that have an implied "you" as the subject.
» **Parallelism:** Keep words and phrases consistent. For example, "This installation requires a hammer, pliers, and a flathead screwdriver" and not, "This installation requires a hammer, pliers, and you'll need a flathead screwdriver, too."

Clarity and Accuracy

Along with using language that is standard and consistent, you need to keep your writing clear and accurate. Writing with clarity is about writing with purpose. Clarity also means that the content is easy to read and easy to understand. To accomplish this, describe what users need to do precisely and when they need to do it using language that is specific. As you'll see in this chapter, pairing text with clear visuals can help, but the writing must first be understandable. For example, don't just say, "Use a computer to go to our website." Be clear and give the exact URL of the web page you want your user to visit.

See Chapter **1** for more on clarity and accuracy.

The accuracy of your language is also important. Usability testing, a process discussed later in this chapter, ensures that your word choice is accurate. If your user needs a screwdriver, for example, what size and type? If you're describing how to complete a keyboard shortcut in a computer program, will it work on both Windows-based computers and Macs? Anticipate the user's unique situation when analyzing your document for accuracy. Recognize that your experience with the task may not reflect the experience of the user.

Using plain language is one way to make your instructions accessible to the largest possible audience. The federal government and many state agencies have adopted plain language policies that promote the use of clear communication. The Plain Writing Act of 2010 defines plain language as "writing that is clear, concise, and well-organized." Clear writing is not only more understandable; it can also save time and money. It takes practice to write this way, however, as this document adapted from the U.S. Census Bureau shows (figure 9.5).

Figure 9.5. Plain Language Guidelines. Many government agencies are required to use plain language in their communications. These steps are useful for any kind of technical writing. Adapted from the U.S. Census Bureau.

1 **Know your audience:** Why does your audience need to read the document?

2 **Organize your thoughts:** What questions will your audience ask?

3 **Summarize main points:** Have you highlighted your main points with headers and lists?

4 **Write short sentences and paragraphs:** Does each paragraph have a topic sentence? Are sentences 20 words or fewer?

5 **Use everyday phrases and words:** Do you avoid unnecessary words?

6 **Minimize jargon:** Have you avoided acronyms, abbreviations, and bureaucratic language?

7 **Use strong subjects and verbs:** Are the subject and verb close together to avoid confusion?

8 **Define uncommon terms:** Are uncommon terms defined and used consistently?

9 **Use headings, lists, and tables:** Can your audience find the material they want quickly?

10 **Proofread:** Has someone else reviewed your document with a fresh pair of eyes?

Scannable Design

The last time you read instructions, did you read them from beginning to end like a chapter in a novel? Of course not. Reading instructions happens in small steps that accompany each stage of the task. Users move back and forth between the instructions and the actual task. For this reason, instructions need to be broken up into sections that can be located easily by scanning the document.

See Chapter 1 for more on scannability.

A scannable design is always important in technical communication, but this factor is particularly important when creating instructions. Notice how you can scan this flyer about the DASH Eating Plan to quickly identify its tips for reducing salt and sodium (figure 9.6). While this document does not include a set of instructions, it is an instructional document that uses similar techniques.

Figure 9.6. Enhanced Readability with Headings. Effective use of images and headings allows helpful information to fit this single page. Source: National Institutes of Health.

The large font makes it clear where the document starts.

This box is the same height as the heading box to create parallelism.

The images create interest but do not overwhelm the content.

Headings are one way to make a document more scannable. Headings show the structure of the document and come in three general categories: question headings, statement headings, and topic headings. The more specific your headings are, the easier it is for users to scan for the information they need (figure 9.7). Question headings tend to be more conversational, while topic headings tend to be more formal. It's up to you to decide which kind of heading is best for your audience and the purpose of your document.

Heading titles should be chosen carefully. Think about the keywords or phrases users will be looking for when they scan the document. Guide the user by providing similar words and phrases from one heading to the next, as in these headings for fence construction: "building your fence," "installing your fence," and "fixing your fence." Keep the headings as short as possible while still being specific.

Figure 9.7. Options for Headings. This example includes three kinds of headings: question, statement, and topic. Notice how the level of formality changes slightly with each heading.

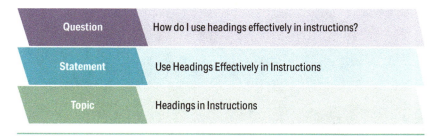

Question	How do I use headings effectively in instructions?
Statement	Use Headings Effectively in Instructions
Topic	Headings in Instructions

Headings are only useful when they are distinct from the rest of the document. That often means using a different style font, a different font size, or white space to draw attention to the heading. Here are some important design techniques to increase the scannability of your document:

» Use plenty of white space.
» Emphasize key ideas or warnings with color, icons, or all caps.
» Enlarge and make clear headlines with your headings.
» Use bold text for important terms.
» Format and structure the document consistently.

Structure and Format

Instructions are often structured vertically as numbered lists. While this is true for many instructions, there are variations. Some variations you might see include fixed-order steps, variable-order steps, alternate steps, and nested steps.

Fixed-order steps require that each step be performed in a sequential order. This technique uses numbered steps. For example, the fixed-order instructions for how to drive a car include the following steps:

1. Put the key in the ignition.
2. Turn the key until the engine starts.
3. Press the foot brake.
4. Put the gear shift into drive.

Variable-order steps are more flexible and can be performed in any order. This technique uses bullet points. Figure 9.6 is a good example of this structure. It doesn't matter what order you complete the suggested steps—they are all meant to help lower sodium. You can reduce your sodium, eat your veggies, or add spice in any order.

Alternate steps include two or more ways of completing steps. In this format, you introduce the idea of alternate options in the introduction. You can use the word "or" to indicate the options within the steps. For example, "Snap the machine lid closed by pressing firmly *or* use the rubber mallet to close the machine lid."

Nested steps are used when the steps are too complex and need to be broken down into substeps for clarification. You indent the substeps and order them as shown in figure 9.8.

Figure 9.8. Formats for Nested Instructions. This table shows the basic structure for nested step instructions with specific examples.

Format Options	Examples

1) List the step
   a) list the substep
   b) list the substep
	Step 4: How to Change a Tire
	 4) Remove the wheel
	   a) Use the lug wrench to remove the lug nuts
	   b) After all the lug nuts are removed, pull the wheel away from the vehicle
	
	 Tip: The order you remove the lug nuts in does not matter.

1. List the step
   1.1 list the substep
   1.2 list the substep
	Step 4: How to Change a Tire
	 4. Remove the wheel
	   4.1 Use the lug wrench to remove the lug nuts
	   4.2 After all the lug nuts are removed, pull the wheel away from the vehicle
	
	 Tip: The order you remove the lug nuts in does not matter.

Effective Use of Visuals

Visuals can improve a set of instructions and provide users with extra guidance for a complex process. Look back at figure 9.6 and notice how the visual elements help organize the information onto a single page. In effective technical documents, visual elements support the text, and the text reinforces the visuals. When using visuals, make sure the relationship between the text and image is clear. A visual should not distract from the content. Instead it should enhance the user's understanding. Your user should be able to quickly glance at the visual and be able to understand what you want them to do.

Visuals can also create concise documents. As the saying goes, a picture is worth a thousand words. Graphics, diagrams, illustrations, and charts can show plenty of information all at once and reduce the need for written description. You can see this principle at work in the before and after example from the National Highway Traffic Safety Administration (figure 9.9).

Figure 9.9. Enhanced Readability with Visuals. Images can communicate complex ideas quickly. Source: National Highway Traffic Safety Administration and Plainlanguage.gov.

BEFORE

This is a multipurpose passenger vehicle which will handle and maneuver differently than an ordinary passenger car, in driving conditions which may occur on streets and highways and off road. As with other vehicles of this type, if you make sharp turns or abrupt maneuvers, the vehicle may roll over or may go out of control and crash. You should read driving guidelines and instructions in the Owner's Manual, and WEAR YOUR SEATBELT AT ALL TIMES.

AFTER

⚠ **WARNING:** HIGHER ROLLOVER RISK

Avoid Abrupt Maneuvers and Excessive Speed.

Always Buckle Up.

See Owner's Manual For Further Information

Another benefit of using visuals is recognition. Even if users aren't familiar with a term (such as a specific tool or part), they might recognize a visual representation. Also, some tasks are so complex that trying to write the steps out (or read such a description) can be challenging. For example, look at the exploded diagram in figure 9.10. An **exploded diagram** is an illustration that allows the user to see all the parts and how they fit together.

Visuals show how the steps connect to each other and fit into a larger process. When thinking about how to use visuals in your instructions, consider the following questions:

» Which type of visual will work best for your instructions?
» Which color theme will you choose for your visuals?
» How much white space is needed to make them easier to read?

A consistent theme and style make your instructions uniform, as when you use similar formatting and color schemes for both tables and figures.

Many technical communicators borrow visuals rather than create them. This can save time, but make sure you follow your industry's guidelines for acknowledging other's work. Just because you find the perfect visual for your document online doesn't mean you have the right to use it. If the owner of the image has given permission for its use, either in writing or with an open license, be sure to recognize the source either with a credit line beneath the image or at the end of your document.

See Chapter ❷ for more on copyright and Creative Commons licenses.

Figure 9.10. Instructions with a Diagram. This page from an assembly manual uses several strategies for organizing information, including an exploded diagram that shows where all the parts will eventually go and a table of parts listed in alphabetical order. Courtesy of Sauder.

This chart visually organizes the many parts required to build this dresser.

Notice how this exploded diagram is labeled so the user can identify each part.

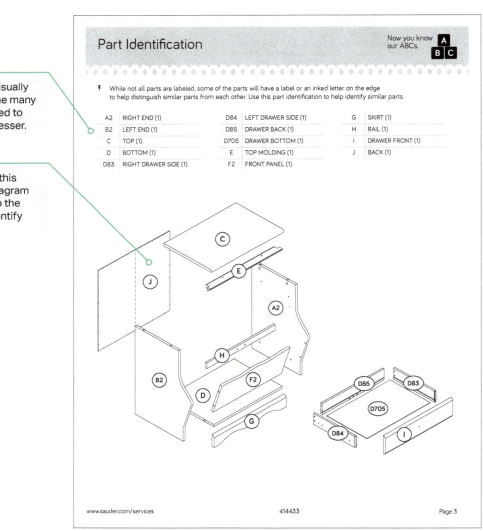

Part Identification

Now you know our ABCs. A B C

While not all parts are labeled, some of the parts will have a label or an inked letter on the edge to help distinguish similar parts from each other. Use this part identification to help identify similar parts.

A2	RIGHT END (1)	D84	LEFT DRAWER SIDE (1)	G	SKIRT (1)		
B2	LEFT END (1)	D85	DRAWER BACK (1)	H	RAIL (1)		
C	TOP (1)	D705	DRAWER BOTTOM (1)	I	DRAWER FRONT (1)		
D	BOTTOM (1)	E	TOP MOLDING (1)	J	BACK (1)		
D83	RIGHT DRAWER SIDE (1)	F2	FRONT PANEL (1)				

www.sauder.com/services   414433   Page 3

Procedures

Procedures establish rules or methods for complex processes. These typically require approval by an authority, such as a department head at a company. Most of the time, companies develop procedures to streamline or standardize a process. For example, a business might give new cashiers a cash handling procedure, as you saw back in figure 9.2. This document makes sure that the training is consistent and accessible, regardless of who gives the training.

Guiding Complex Situations

Procedures help users make choices as they move through complex situations rather than telling them exactly what to do step by step. Procedures often involve multiple decisions, phases, and sometimes even multiple instruction sets. Effective procedures depend on how well you guide users through the decision-making process.

Businesses often create procedures to ensure consistency among employees. For example, a retail business wants to train its managers to deal with difficult customers. A set of instructions would be impractical; no instruction set could anticipate the complexity of dealing with every situation involving a difficult customer. Procedures, on the other hand, give employees guidance to make the best decision in those situations by providing them with multiple options.

Purpose

Companies create procedural documents for a variety of reasons. Here are some examples:

» How to use a company vehicle
» How to seek grant funding
» How to write appropriate workplace emails
» How to apply for family leave

Notice that the topics in this list couldn't be dealt with in a set of instructions. Instead, a procedure needs to outline the best way to make decisions about these topics. Remember, procedures are typically about the roles and responsibilities of people responding to a complex situation.

Audience

When thinking about the audience for a procedure, consider the conditions and the impact of the process. Here are some questions to ask:

» Who will complete the practices, and who will approve them?
» What will the steps cost in money and effort?
» Where will procedures intersect with safety and productivity?
» When will a new procedure be most helpful?
» Why does the current procedure work or not work?
» How will procedures be updated?

Success depends on your keeping procedures current, uniform in design, and simple enough for anyone to understand.

Best Practices for Procedures

A few key principles can make a big difference in the effectiveness of your procedures. As you write, think about the importance of user context, the key steps of the procedure, and useful design.

User Context

Procedures deal with the big picture, so it's important to give the user an understanding of the context. Often, a procedure is created in response to past events. For example, a business that has dealt with embezzlement by a former employee might create a hiring procedure that involves a background check. To help future managers understand the purpose of this procedure, the document should explain that the policy of background checks stems from the possibility of employee misconduct. This background information is the context for the procedure and part of what makes it relevant to the user.

Procedures should provide clear definitions so that users know who should be involved and why. For instance, notice how the document in figure 9.11 created for a lab is specific about who is qualified to carry out the procedure and the required training.

Figure 9.11. Standard Operating Procedure Example. SOPs create a reference for how to complete a complex yet routine task.

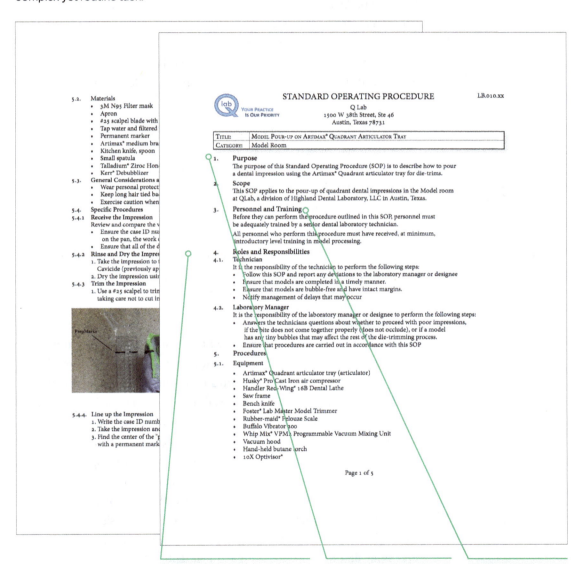

Notice that much of this procedure looks like a set of instructions. This is common when part of a procedure involves a detailed, specific task.

"Purpose" describes what the procedure aims to do. "Scope" defines the specific use for this document.

The SOP defines who is qualified to carry out this task.

Key Steps

A typical procedure includes a description of key steps. These steps are the actions required by the process in order for it to be completed properly. For example, refer to the example of the hiring procedure in "User Context" that requires a background check mentioned in the previous section. An overview of the key steps for this procedure might look like this:

» Review job applications.
» Conduct hiring interview.
» Perform background checks.
» Make hiring offer to prospective employee.
» Complete hiring paperwork.
» Schedule employee's start day.
» Conduct employee orientation.

These steps may look like instructions, but they are more of an outline that guides the user toward the end result. Within each step is a number of decisions that are not spelled out.

Useful Design

As with instructions, procedures need thoughtful design, including clear organization, headings, and visuals. Flowcharts are a good example of a visual that is often incorporated into a procedure. Often, a busy employee can determine the correct course of action simply by referring to a quality flowchart. As the creator of a procedure, review the design of your document to make sure the procedure is easy to understand and follow.

The design of the procedure should help the user navigate it quickly. The following model shows the table of contents for a standard operating procedure for NASA (figure 9.12). Note how this document is structured using the nested format. A functional and consistent design is an important component of procedures, especially if they involve the level of technical skills described in this sample document.

Figure 9.12. Standard Operating Procedure Overview. NASA uses best practices for the organization and design of its SOPs. Source: NASA.

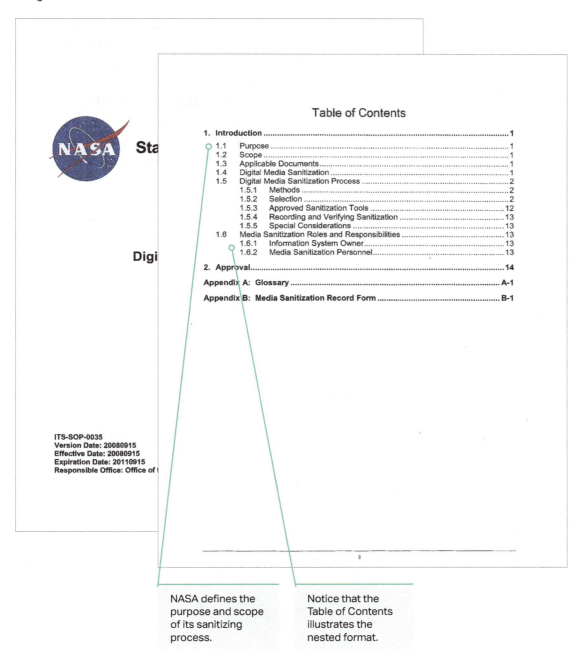

NASA defines the purpose and scope of its sanitizing process.

Notice that the Table of Contents illustrates the nested format.

Legal and Ethical Concerns

With both instructions and procedures, it's important to think about potential liabilities. Instruction sets, for example, might require tools that could cause harm if used improperly. The writer of instructions can't assume that the user will know how to use a tool safely. For this reason, many instruction sets include warnings (figure 9.13). Warnings should be given before the user begins the task, and the warning should be accompanied by some kind of visual marker, such as an icon, contrasting color, white space, or large text to set it apart from the rest of the document.

Always consider safety when writing instructions. Do you need to remind the user about wearing eye protection or ensuring their workspace has proper ventilation? Some instruction sets include broad statements about the user taking full responsibility for any activities involved in the instructions. Technical communicators should consult with their company's legal department to discuss how to approach possible liability.

Unclear wording can be a problem with instructions or procedures. If a user misunderstands instructions due to poor wording, the result could be damage to property or bodily harm. If a procedure is unclear due to language, the result could be the loss of company profits, damage to a business's reputation, or employee dissatisfaction. Technical communicators are responsible for testing instructions and procedures by submitting them for review. This means you must budget additional time to allow for conducting usability tests and revising the document based on those results.

Usability Testing

Usability testing allows the technical communicator to test a document to determine its effectiveness. Any technical document can benefit from usability testing, but it's especially important in the case of instructions and procedures. If a technical document does not meet the needs of its user, or only half meets those needs, there can be serious consequences.

Figure 9.13. Instructions with a Warning. This assembly manual places warnings and important information prominently on the front cover. Courtesy of Sauder.

Notice the placement of the caution symbol to draw attention to this warning.

The manufacturer provides contact information in case the user has any trouble with the instructions.

During the usability testing stage, technical communicators try out their document in a low-risk environment to catch errors. Don't assume that just because something makes sense to you, it will make sense to everyone. Even highly experienced technical communicators need to test their instructions and procedures with potential users.

The U.S. Department of Health and Human Services offers suggestions for usability testing on its website, www.usability.gov, a site designed to help students and practitioners focus on the user experience in government and the private sector. Their site sorts testing into four categories: concurrent think aloud, retrospective think aloud, concurrent probing, or retrospective probing. Each method obtains feedback through a slightly different means, and each has positives and drawbacks to consider before you decide which one to use (figure 9.14).

Best Practices for Usability Testing

The following best practices can help make your usability test more successful. These guidelines will help your test maintain objectivity, accessibility, flexibility, and repeatability.

Avoid Anchoring

Beware of influencing your test-takers. The psychological term **anchoring** refers to the way outside questions and comments can influence the thinking and responses of study participants. This is a common pitfall in usability testing. When handing your document to a tester for review, don't explain: "So, what I'm trying to do here is…" or "The point of this document is to…." With just one sentence, you have biased their views, changed their understanding, and disrupted the possibility of actual feedback. Resist this temptation. You won't be there when users are accessing your document in the future.

Figure 9.14. Usability Testing Pros and Cons. This table provides pros and cons for the four categories of usability testing. Source: U.S. Department of Health and Human Services.

Pros	Technique	Cons
Understand participants' thoughts as they occur and as they attempt to work through issues they encounter Elicit real-time feedback and emotional responses	Concurrent Think Aloud (CTA)	Can interfere with usability metrics such as accuracy and time on task
Does not interfere with usability mechanics	Retrospective Think Aloud (RTA)	Overall session length increases Difficulty remembering thoughts from up to an hour before = poor data
Understanding participants' thoughts as they attempt to work through a task	Concurrent Probing (CP)	Interferes with natural thought process and progression that participants would make on their own if uninterrupted
Does not interfere with usability metrics	Retrospective Probing (RP)	Difficulty in remembering = poor data

Design for Accessibility

Accessibility refers to the practice of creating content and experiences for a wide range of people, including individuals who may have temporary or permanent visual, motor, auditory, speech, or cognitive disabilities. Designing for accessibility also means including people who speak multiple languages or who may be from different cultures. Inclusive technical documents that reduce barriers to understanding ultimately improve the experience for all users.

If you are designing a technical document for people with disabilities, you should include people with disabilities in the testing phase. Consider testing the document with people who have limited access, such as visually impaired participants or people who do not own a computer. If you are designing a document for English language learners, then test it on users who do not speak English as their first language.

Accessible design involves being conscious of the way your text and images work together. Keep in mind that color blindness, low vision, hearing impairments, learning disabilities, and any number of temporary disabilities may impact how the user interacts with your document. Test with participants from diverse backgrounds, cultures, and experiences to ensure inclusivity. Testers are usually a small group, so make sure they are as representative as possible.

Be Flexible

Be prepared to adjust your usability test and the document itself in response to feedback from your test group. Testers will do things you don't expect. This is because they didn't design the document. You need this feedback to improve your document, even if it's unexpected.

Repeat as Needed

When you set deadlines for your projects, make sure that you include time for multiple rounds of usability testing. The best development process has multiple versions of a document, as well as multiple tests.

Considerations in Usability Testing

Usability testing should examine the ease of performance, efficiency of performance, degree of error, and aesthetic quality as users interact with your document. Technical communicators need to listen and watch carefully to understand the audience's needs and where the document may send them in the wrong direction. Before you begin, consider how you will determine these four principles:

» **Ease of performance:** How easily can users accomplish what is needed on their first use of your document? How intuitive is your design? How accessible is your content?

» **Efficiency of performance:** How quickly can users accomplish what is needed as they interact with your document?

» **Degree of error:** How often and how many errors do users make during the testing? How severe are the errors? What kind of recovery do they make after their errors?

» **Aesthetic quality:** How pleasing is the design to users? Is encountering your document a positive experience? Are there aspects that are off-putting?

Preparing a Usability Test

Usability testing requires you to keep track of many factors, so keep these questions in mind as you prepare to test your document:

Plan the Test

» **Identify scope and purpose:** How detailed is the test? What are you trying to determine?
» **Figure out logistics:** When and where will the test happen? How long will it be? What equipment (if any) is necessary?
» **Plan scenarios:** Do you need to design a situation in which to test the document?
» **Determine metrics:** How will you measure the reaction of test-takers? How will you collect feedback?

Recruit for and Give the Test

» **Research:** Who is your target participant?
» **Recruit:** How will you find willing test-takers who accurately represent your target user?
» **Moderate:** How will you administer the test?

Assess the Results

» **Evaluate:** Does anything in the document need to change based on the test results?
» **Refine document:** What changes in the document will produce better results?
» **Repeat testing:** What happens when the refined document is tested again?

Case Study

Instructions and Procedures at Work

This case study is an opportunity for you to put into practice what you've learned. Look at the following case study to consider how you would respond using best practices introduced in this chapter for both instructions and procedures:

Rosario is the safety manager at Wrecking Ball Demolitions. This role puts Rosario in charge of reducing the number of injuries, co-planning safe work practices with crew leaders in the field, and ensuring that the company's safety equipment is complete and in good condition. In addition, Rosario occasionally documents how employees should complete certain tasks to ensure a safe work environment.

Recently, Rosario noted an increase in reported injuries involving reciprocating saws in the field. When she examined the incident reports, she realized that employees weren't following basic safety procedures or reporting their injuries in a standardized fashion.

To address this problem, Rosario replaced faded safety notices and relocated them near the equipment. She also wrote a standard procedure for yearly equipment training that includes a process for reporting injuries. Her solution was successful because Rosario understood the purpose (to reduce injury and standardize reporting) and the audience (her employees) for her documents. Her understanding of the fundamentals of technical communication allowed her to create a clear message in the form of a new reporting procedure and more visible and useful safety procedures.

Discussion

» For both instructions and procedures, awareness of the audience is crucial. If you were in a situation like Rosario's, how would you balance the technical nature of your document with what you know about your users?

» How might Rosario approach this scenario differently if she were documenting the procedures for reporting injuries for HR?

» What are some best practices that Rosario could incorporate into her response?

Checklist for Document Accessibility

Color Use

- ☐ Avoid using color as the only means of representation.
- ☐ Ensure sufficient contrast between the background and text.
- ☐ Consider cultural connotations and common usages of colors.

Headings

- ☐ Use the application's built-in Styles tool for headings.
- ☐ Follow a logical nesting order and do not exceed six heading levels.

Formatting and Layout

- ☐ Avoid underlined text for emphasis as it can be mistaken for a link; use bold and italics instead.
- ☐ Avoid using tables for layout.
- ☐ Check document for consistency in style and navigation.

Images

- ☐ Provide alternative text (alt-text) that conveys the same information as the image itself.
- ☐ Ensure that images can be enlarged to 200% without pixelating.

Conclusion

Once more, think back to your first job and the uncertainty that you might have felt. You needed information and the confidence that you could complete new tasks well. The situation is similar for users of instructions and procedures. By creating accurate, clear instructions, you make it possible for your audience to complete their tasks safely and efficiently. By developing procedures that answer potential questions and direct decision-making, you can save time and effort.

Remember that your task is to understand the situation of your audience and help them with information that is reliable and easy to understand. Give users a simple and well-designed document so they can complete their task with ease. When creating a procedure, demonstrate that you understand why the procedure is needed and how it will be applied so that users will use it with confidence. By anticipating needs, testing your documents, and revising to make your communication as useful as possible, you can save users frustration and enable them to succeed.

Chapter 10

Proposals and Short Reports

Abstract: A technical proposal involves making a persuasive argument supported by research for an idea that needs approval to move forward. Technical proposals are a type of short report and can be written for internal or external audiences. Proposals may be requested by a company (solicited) or offered by an individual to a company (unsolicited). The various proposal types require the technical communicator to carefully consider the document's purpose, message, and audience. Beyond the proposal, short reports cover all kinds of technical situations. They provide information, analysis, or recommendations to solve problems. They monitor or document progress, clarify policies, and guide change. Proposals and short reports require succinct, specific, objective, and ethical communication. Ultimately, these versatile documents bring together many key principles of technical communication.

Looking Ahead

1. Why Proposals and Short Reports Matter
2. Proposals Defined
3. Types of Proposals
4. Typical Elements of Proposals
5. Short Reports Defined
6. Steps for Writing Proposals and Short Reports
7. Principles for Proposals and Short Reports

Key Terms

- » appendix
- » external proposal
- » Gantt chart
- » internal proposal
- » proposal
- » research report
- » request for proposal (RFP)
- » sales report
- » short report
- » solicited proposal
- » status report
- » unsolicited proposal

Why Proposals and Short Reports Matter

Proposals and short reports are used in most fields, and, as a technical communicator, you'll likely be asked to write one at some point. While proposals and short reports can take on different forms, most include similar sections and have a similar goal—to get approval on a project that needs to move forward.

Many employees have no idea how to create these documents, yet proposals and short reports are two of the most frequently used documents in the professional world. Knowing how to write in these formats will increase your chances of getting what you want and make you a valuable asset in the workplace. Practical experience with proposals and short reports give you an excellent foundation for professional communication.

Proposals Defined

A **proposal** is a technical document written for decision-makers to convince them to choose a specific course of action. It's often the first step to getting a project started. A proposal begins with research on a problem. The proposal's goal is to provide enough information for a decision-maker to greenlight a solution to the problem.

Proposals are often used within an organization to request changes. For example, maybe you'd like a standing desk in your work space. How would you convince your boss to approve the request? Or maybe the proposal is about a significant external issue. For instance, your engineering firm would like to bid on a bridge retrofit for the city. Either way, to write a successful proposal, you must do two things:

» Inform the decision-maker about the issue
» Suggest a plan of action

See Chapter **7** for business email and memo formats.

Shorter proposals are often presented in a business email or memo format with headings. Longer proposals, which can contain as many as ten sections, generally have a cover letter and are submitted as a separate document, either electronically or in print. A basic format like the one included in this chapter can get started (figure 10.1).

Figure 10.1. Basic Proposal Format. Many organizations and businesses have their own proposal format. This basic format can be adapted to fit almost any project.

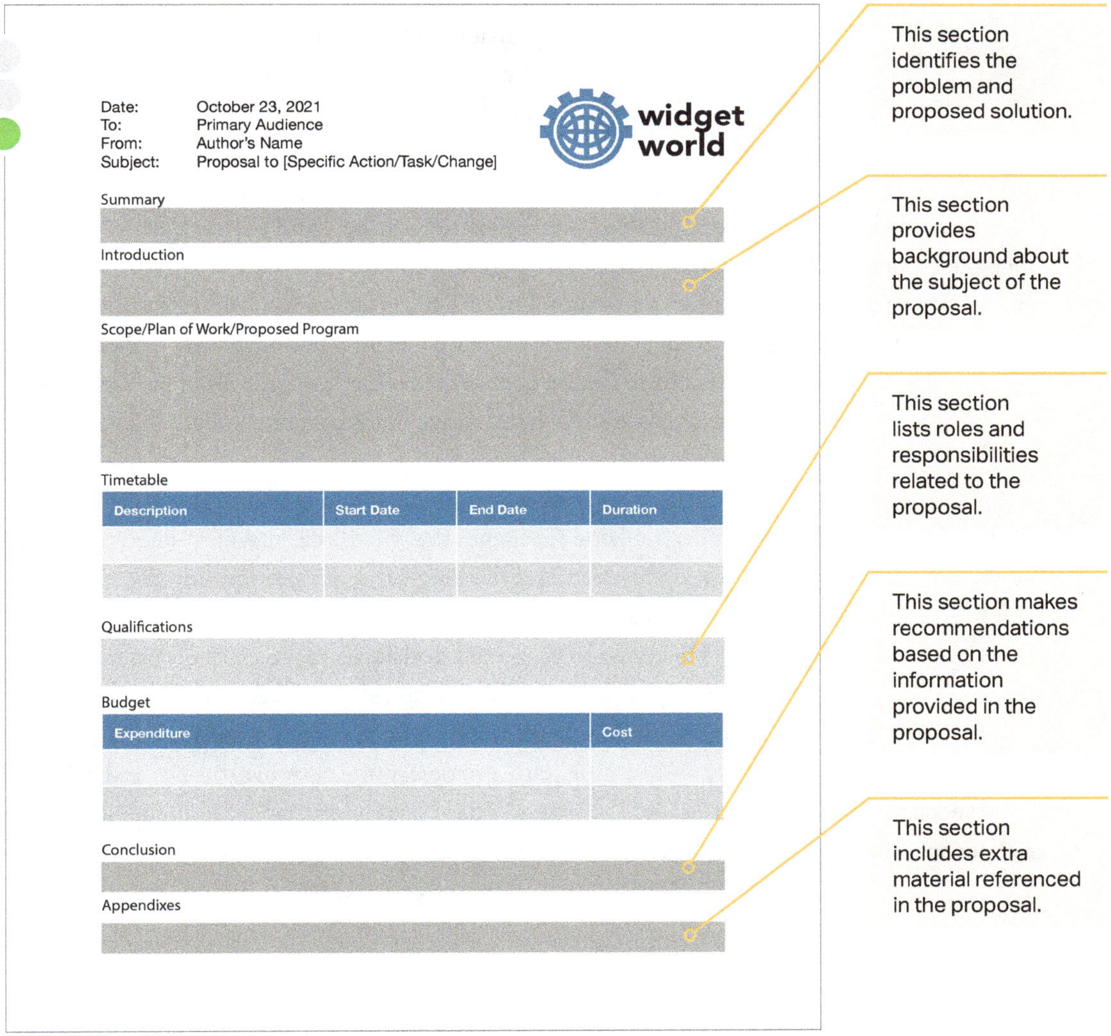

Date: October 23, 2021
To: Primary Audience
From: Author's Name
Subject: Proposal to [Specific Action/Task/Change]

widget world

Summary

This section identifies the problem and proposed solution.

Introduction

This section provides background about the subject of the proposal.

Scope/Plan of Work/Proposed Program

Timetable

Description	Start Date	End Date	Duration

This section lists roles and responsibilities related to the proposal.

Qualifications

Budget

Expenditure	Cost

This section makes recommendations based on the information provided in the proposal.

Conclusion

Appendixes

This section includes extra material referenced in the proposal.

Proposals at Work

Let's take a look at a proposal submitted by an employee tasked with solving computer issues for the company. Roy is the IT manager at Widget World, a midsize computer parts company. Roy's job duties include recommending changes that will benefit the company's bottom line or increase productivity.

Roy discovers that his workplace spends a large amount of money every year buying and maintaining desktop computers. Roy believes he has an answer that could save the company money. He knows that many offices are converting to a virtual machine (VM) system in place of traditional PCs for every employee. The new system, however, requires an investment of cash up front. How will he convince his boss that the expense of the new system is needed?

See Chapter 11 for more on long reports, including feasibility studies.

While writing his proposal, Roy keeps his audience's needs in mind. He suspects his boss will be cautious at best and skeptical at worst, so he collects relevant data to support his proposal. He may gather more than what is needed for the proposal in case his boss wants a full feasibility report, a long report that studies whether a solution is reasonable and beneficial. But Roy also needs to decide how much research to give his boss. The decision of what to include and what to leave out is part of the challenge of shorter reports and proposals.

See Chapter 5 for more on primary and secondary research techniques.

Roy focuses his research on the facts and evidence his audience needs to make a decision. He researches the cost of the current systems, the cost of buying the new system, and the long-term net savings. Roy also researches the current annual cost of buying and maintaining desktop computers at his company. He knows by the time a decision is made, costs of current systems and the new system will likely increase, so he will add those projections to his recommendation. In addition to crunching numbers, Roy surveys his coworkers and administrators to determine current attitudes and habits of computer usage. The blend of hard data and personal opinions will play a big part in whether his proposal is agreeable to his boss.

Proposals and the Problem-Solution Framework

Think back to the Problem-Solution Framework presented in Chapter 1. This concept will help you create a successful document. When writing a proposal, always consider the problem and solution in relation to the specific situation.

In a proposal, the problem takes the form of an issue that needs resolution or an opportunity that could be realized (figure 10.2). After clearly establishing the problem, you must present a solution in the form of a message. To be convincing, your solution must be realistic and rooted in evidence (rather than guesswork). As with all documents, consider how your audience and purpose contribute to your message.

Figure 10.2. Problem-Solution Framework: Short Report. The purpose of a proposal is to point out a problem in the form of an issue or opportunity. An effective proposal shows how the issue can be resolved or the opportunity can be realized.

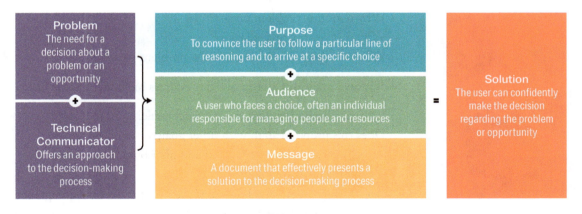

Audience

A successful proposal carefully considers the user's situation and other factors that impact decision-making. The user of a proposal is typically a manager. One common reason a proposal fails is that it does not accurately or honestly recognize the current situation, which includes both the user's and the company's needs at the moment.

Timing is an important part of a proposal. Because of this, we recommend that you work directly with your audience during the research phase whenever possible. Conduct interviews, send emails, or administer surveys to collect information from others to find out if the time is right and the audience is ready to listen. While Roy works on his proposal, he can meet with his boss and get her perspective so he can better address any concerns.

Here are a few questions to determine the direction of a proposal:

» What is the problem being addressed?
» Whom does this problem impact?
» Why does this problem exist?
» Why does this problem persist?
» Why is it worthy of a solution?
» What are the potential hazards the problem presents?

Additional questions a proposal may address include the following:

» Is there a limited budget or a tight schedule?
» Does this proposal align with other goals?
» How will the proposal be shared and distributed?
» Who makes the final decision?
» Do any regional, cultural, or linguistic factors affect the acceptance or implementation of this proposal?

See Chapter ❶ for more on creating a user profile.

In Roy's case, he must keep in mind that his boss, Carmen, is not a technology expert. As a result, Roy must explain the technical side of the proposal in a way that won't frustrate his boss. He needs to use plain language and avoid technical jargon that he and his fellow IT professionals use. Roy also needs to consider that, like most bosses, Carmen is cautious about any expenses associated with the proposal. He needs to make a case for why the project will be worth the associated costs. Even if Carmen agrees with his proposal, she will need to take it to the board, which consists of a multinational group of stakeholders. Roy's proposal must be clear and convincing enough to make Carmen want to take that next step.

Purpose

Your purpose is to provide an evidence-based, objective argument so the user can make an informed decision. This goal will help you identify what information to include as you write your proposal. Decision-makers who read proposals are busy people. Keep your document simple, short, and focused. Avoid the temptation to overload the user with information. This is not a situation where more is better. Instead, be selective and share only the best information.

Roy's purpose is to convince his boss that the hardware upgrade will benefit the company in spite of the up-front expense. He explains this purpose by showing Carmen that the current setup is hurting the company's bottom line by generating unnecessary long-term maintenance costs. He must show that his proposed hardware upgrade is feasible, realistic, and cost effective in the long run.

Message

Your message is shaped by the type of proposal you choose. In the following section, we'll introduce several proposal types. Keep your message in line with the document type. For example, you don't write a planning proposal the same way you write a sales proposal. The intent of your specific proposal type and the needs of the user matter when creating your message.

Roy's proposal was not requested by his boss, so he must convince Carmen that the project is worthwhile. As a result, his tone must be persuasive while also rooted in facts. Roy uses a format similar to a goods and services proposal that includes cost savings resulting from a proposed purchase. While he's not offering a good or service to a customer, his approach will be similar because he must show his boss that the benefits of the project outweigh the investment.

Types of Proposals

There are four main types of proposals: internal/external and solicited/unsolicited (figure 10.3). The specific approach for writing a proposal depends on the topic and desired outcome. Proposals can be about company policy or sales goals, for example. Or they might be written in response to a **request for proposal (RFP),** a document put out by a company that asks for bids on goods, services, or solutions. Proposals might be written by an individual or by an entire department.

Identifying the type of proposal can help you make important decisions about the document's audience, purpose, and message as shown in this table of proposals (figure 10.4).

Figure 10.3. Proposal Categories. This chart shows the four categories of proposals.

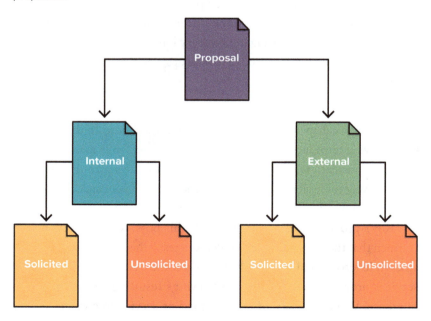

Figure 10.4. Common Types of Proposals. This table lists some of the most common types of proposals and how audience, purpose, and message differ for each.

Type	Audience	Purpose	Message	Example
Grant Proposal	Government agency or nonprofit	To obtain funds	"Support our viable project."	A lab applies for funding to expand its research capacity.
Planning Proposal	Decision-maker who can approve a project	To justify a plan of action	"Here is our offer to bid on or complete a project."	A business proposes that annexing land will benefit the city in spite of potential complications.
Research Proposal	Decision-maker for a business, organization, or company	To provide evidence to support a change	"We want to gather information for a project."	An engineering firm proposes a seismic study for an at-risk structure in response to an RFP.
Sales Proposal	Contractor, business, company, or organization	To offer goods or services	"Buy our product or service."	An insurance company proposes a detailed alternative to another business's existing employee health plan.

Internal / External

Whether the problem is big or small, internal or external, the type of proposal is not always determined by the writer. A semiconductor company that wants to install new machines in its cleanroom may request proposals from more than one company. Or a technology department may be tasked with reducing cost and request proposals from its own employees.

An **internal proposal** happens *inside* a specific business or organization. Roy's proposal is internal because he's writing it for his boss. Some companies will have an established procedure that is required for an internal proposal to be successfully pitched.

An **external proposal** is completed when an *outside* party writes a proposal for a different business or organization, typically as a profitable business transaction. For instance, when the semiconductor company wanted to install new cleanroom machines, they put out an RFP from three different companies before they made a decision.

Solicited / Unsolicited

Another category of proposals involves whether the document is created in response to a direct request.

A **solicited proposal** is specifically requested by the audience (typically a business). In this case, one business has approached another business (or sometimes an individual) about a possible project. The first business may do this by issuing an RFP, asking potential business partners to write and deliver a proposal, as in the example of the semiconductor company. Or a business in need of a new building might issue an RFP, for example, to multiple architects as a way of getting multiple suggestions.

An **unsolicited proposal** is more speculative because the party writing the document is not responding to a direct request. In this case, the business or individual writing the proposal hopes that the proposal will be appealing to the recipient. Most of the time, the business or individual writing an unsolicited proposal has already done a significant amount of research to determine the likelihood of the proposal being accepted, similar to what Roy has done.

Typical Elements of Proposals

Proposals tend to follow a similar structure, which typically includes several (if not all) of the following sections.

Summary

A shorter proposal might not need a summary, but longer proposals of more than a few pages will. Summaries are also called executive summaries or abstracts in some cases. Often this section is included on the title page, and sometimes it is limited to a specific word count. The summary's purpose is to provide a quick rundown of the proposal, as you can see here with Roy's summary (figure 10.5).

In some cases, a user may review only the summary before deciding if they want to read the entire proposal. As a result, it's wise to think of a summary as a short sales pitch.

When writing a proposal summary, present the key information in an abbreviated format. This section should briefly and clearly state the proposal's problem and provide a brief overview of your recommendation. If the decision-makers find something useful in the summary, they'll have a reason to keep reading. The abstracts at the beginning of each chapter in this textbook function in a similar way.

Figure 10.5. Proposal Summary. This model shows the opening of Roy's proposal to his boss.

Date: October 23, 2021

To: Carmen Lopez

From: Roy Moss

Subject: Proposal for IT Department Cost Savings/VM Integration

Summary

The current cost of replacement and maintenance for physical hardware could be reduced. We spend over 40 percent of the department's budget every year purchasing new desktops and servers and additional money maintaining the existing physical hardware. The IT department seeks approval to convert to a virtual machine (VM) system that will reduce costs due to the elimination of hardware replacement and employee time spent in maintenance. The conversion will only take two months. While the initial cost of moving to the new system would be around $8,000, the estimated annual savings for the company will be around $4,000.

Roy is specific and shows what percentage of the department budget is allocated to this type of expense.

The projected cost-savings of this change is what people really want to know.

Introduction

The introduction sets the stage for the body of the proposal. It's where you provide background, an overview of key ideas in the proposal, and a preview of the document's organization.

The introduction is different from a summary. While the summary condenses the entire proposal into a smaller package, an introduction smoothly leads into the main content of the document. In Roy's introduction, he links his proposal with the company's "ongoing goal of reducing technology expenses" in order to show how this idea aligns with other priorities (figure 10.6).

Figure 10.6. Proposal Introduction. This model shows the introduction of Roy's proposal to his boss.

Roy uses targeted language and avoids personal opinion.

Using evidence is more convincing than vague statements.

Introduction

In keeping with the ongoing goal of reducing technology expenses at Widget World, I've been examining our physical hardware costs. It turns out we spend over 40 percent of the department's budget every year purchasing and maintaining physical hardware. We spend an additional 25 percent of the budget on desktop computers and servers every year. Research shows that we could operate with fewer physical servers and maintain productivity. In addition, we could replace our physical desktop hardware with alternative equipment to save additional money.

Scope

The scope of a proposal explains the suggested course of action. It is the heart of the proposal. In this section, you must provide convincing evidence that your recommendation is the best choice. Your evidence comes from talking to experts, consulting appropriate sources, and possibly doing original research or experimentation. This section is sometimes referred to as a proposed program or a plan of work. As you can see in figure 10.7, Roy's scope includes multiple steps.

See Chapter 5 for more on primary and secondary sources.

Figure 10.7. Proposal Scope. This model shows the scope of Roy's proposal.

3. Tran

We'll al
going. T
transfer
good ne
freed by

4. Cha

We can
the data
away. T
desktop
time to

5. Repl

After we
can get
device

Thin Client

Figure 1. "Thin

Scope

This proposal applies to the elimination of most desktop computers at Widget World and replacing them with a VM system. This proposal also includes the recommendation to convert our existing servers to VM hosts. This proposal does not include any changes to laptops or other electronic equipment maintained by the company.

1. Train staff

We'll need to train the IT staff on the process of switching to the VM system. None of us have done this before, so it will take about a week for us to learn the process. We can take training for a modest fee, or we can try to teach ourselves, which will save the cost of the training, but take more hours. I suggest the paid training, which will only take a day. We could even do it on a Saturday to save work hours during the week if the company is willing to pay for the overtime.

2. Change current servers

We can consolidate the large number of servers into a smaller number by making some of them VM servers. VM servers will take less space, saving us the cost of maintaining and replacing the number of servers we currently have. Just three physical servers converted into VMs will take the place of our old setup with six physical servers. The extra physical servers that we free up could then be used to host PCs, allowing us to eliminate the desktops in Step 4.

Timetable

A well-researched time frame gives the audience a realistic sense of how long the project will take. This section is important for decision-makers. An accurate timetable that is easy to interpret will make your proposal more valuable. Roy uses a Gantt chart to help his boss see the various stages of this project (figure 10.8).

Gantt charts are used to chart out tasks and timelines in collaborative projects where multiple assignments need to be completed at different times to achieve a specific outcome. The chart defines expectations and timelines, which keeps the team accountable and efficient.

Figure 10.8. Project Timeline. This model shows the timeline of Roy's proposal to his boss using a Gantt chart.

The Gantt chart, also known as a bar graph, is a useful visual to show the timeline of multiple stages of a project at once.

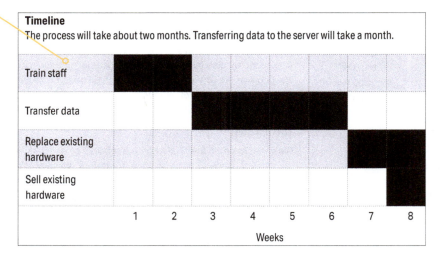

Timeline
The process will take about two months. Transferring data to the server will take a month.

	1	2	3	4	5	6	7	8
Train staff								
Transfer data								
Replace existing hardware								
Sell existing hardware								

Weeks

Qualifications and Experience

This section of your proposal outlines your qualifications and experience to convince the audience that you have a reasonable plan for solving the problem. Making a clear proposal backed by solid research is a crucial part of showing that you have the ability to carry out the proposal successfully.

To begin, ask yourself if you have specialized knowledge or skills that make you the ideal candidate for this work. If so, state your experience explicitly. For example, Roy has a track record of problem solving in his department. He is also an expert in his field. If part of the plan involves experts in other areas, say so. The more transparent your proposal is, the better chance it has of being approved.

Budget

Decision-makers care about the cost of undertaking new projects. The budget section needs to provide realistic expense estimates (figure 10.9). As always, the numbers you use here need to be based on accurate research. Roy gathers quotes from other companies so he can provide associated costs of current and new systems. He also includes associated implementation costs to get a true idea of what the plan will cost the company.

Figure 10.9. Proposal Budget. Proposals should include any potential costs. This model shows Roy's project estimates.

Source of Project Cost

	PROJECT TASKS	LABOR HOURS	LABOR COST ($)	MATERIAL COST ($)	TRAVEL COST ($)	OTHER COST ($)	TOTAL PER TASK
PROJECT DESIGN	Develop Functional Specifications	120.0	$150.00	$75.00	$200.00	$75.00	$620.00
	Develop System Architecture	150.0	$100.00	$100.00	$200.00	$100.00	$650.00
	Develop Preliminary Design Specificatio	150.0	$100.00	$150.00	$200.00	$100.00	$700.00
	Develop Detailed Design Specifications	100.0	$150.00	$50.00	$200.00	$75.00	$575.00
	Develop Acceptance Test Plan	150.0	$200.00	$50.00	$200.00	$75.00	$675.00
	Subtotal	**670.0**	**$700.00**	**$425.00**	**$1,000.00**	**$425.00**	**$3,220.00**
PROJECT DEVELOPMENT	Develop Components	120.0	$150.00	$75.00	$200.00	$75.00	$620.00
	Procure Software	150.0	$100.00	$100.00	$200.00	$100.00	$650.00
	Procure Hardware	150.0	$100.00	$150.00	$200.00	$100.00	$700.00
	Development Acceptance Test Package	100.0	$150.00	$50.00	$200.00	$75.00	$575.00
	Perform Unit/Integration Test	150.0	$200.00	$50.00	$200.00	$75.00	$675.00
	Subtotal	**670.0**	**$700.00**	**$425.00**	**$1,000.00**	**$425.00**	**$3,220.00**

Source of Project Cost | Expenditures Over Time | Cumulative Project Costs | Data Worksheet | +

Conclusion

A conclusion smoothly brings the document to an end with a clear transition and a strong final statement. Be sure to review the purpose of the proposal. In other words, remind the audience of what you are asking them to approve. Always recap the main points of the proposal in a short format. Leave the audience with something to think about in your closing statement that highlights the lasting benefits of the proposed solution.

Appendixes

This section isn't mandatory, but it is common in many proposals and reports. An **appendix** collects additional information at the end of a proposal or report and may include raw data, calculations, graphs, and other materials that were part of the research but not essential to the proposal itself. For proposals and other short reports with a lot of background information, there may be more than one appendix. Refer to each appendix at the appropriate point (or points) in the body of your proposal by inserting a parenthetical citation with the relevant appendix at the end of the sentence. If you have more than one appendix, assign each appendix a capital letter, as in Appendix A, Appendix B, Appendix C, etc.

Short Reports Defined

See Chapter 11 for more about formal reports.

Short reports take many forms, but they are always brief documents that provide information about a specific objective, event, or ongoing issue. When you write short reports, your goal is to inform your audience clearly and simply. You should save your audience time and effort by considering their needs. What do they need to know? What is unnecessary?

In many ways, short reports are similar to formal reports. The difference lies in the level of detail or complexity. While formal reports tend to be long, contain multiple sections, and frequently involve significant expense for a business, a short report deals with a single issue that may be less complex.

Short Reports and the Problem-Solution Framework

As you can see in the Problem-Solution Framework, the problem in a short report begins with an issue that needs resolution (figure 10.10). Consider the amount and detail of information the user needs. Present the information in the short report to fit the exact needs of your audience. The information could be data, description, evaluation, or another mode that aids understanding. Your success in writing a short report depends on how easily and accurately users get what they need.

Figure 10.10. Problem-Solution Framework: Roy's Report. The Problem-Solution Framework keeps Roy on track as he works to solve the problem.

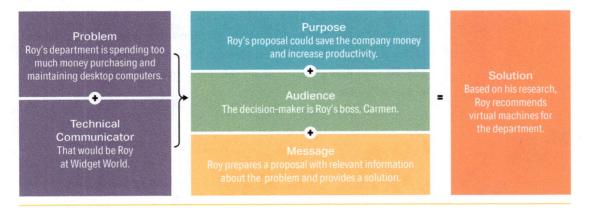

Problem
Roy's department is spending too much money purchasing and maintaining desktop computers.

+

Technical Communicator
That would be Roy at Widget World.

Purpose
Roy's proposal could save the company money and increase productivity.

+

Audience
The decision-maker is Roy's boss, Carmen.

+

Message
Roy prepares a proposal with relevant information about the problem and provides a solution.

=

Solution
Based on his research, Roy recommends virtual machines for the department.

Audience

The audience for a short report is typically a manager, such as a department supervisor or the head of a business or organization. Respect their time by keeping your reports focused and as brief as possible. Remember that a decision-maker will likely approach your document with a critical mindset. They won't necessarily take your word for it when you make a claim—your job is to convince them. You can do this by anticipating and addressing the questions that they'll most likely ask.

In Roy's case, his document is a status report for his boss. Carmen cares about the success of the project and wants an update that she can review quickly to understand how close the project is to completion. Armed with this information, Roy can create a report that meets his boss's needs.

Purpose

When writing short reports, your purpose is to inform the user about specific data, decisions, or policies, among other possibilities. The success of your short report depends on how easily and accurately the audience understands the content. Be clear and eliminate anything that distracts from the key details. Be objective and represent information accurately.

After getting Carmen's approval, Roy drafts a status report with the goal of providing accurate and timely information. This means that he must give updates regardless of whether the project goes according to plan. Sometimes, status reports must relate bad news. If the project falls behind schedule, for example, Roy needs to explain why in the report. Typically, a status report also offers solutions to problems that happen along the way. If Roy discovers an unexpected expense, he needs to report this to his boss. Ideally, he'll find a way to offset this extra cost and account for it in his report as well.

Message

The message in a short report should match an expected format. Decision-makers are familiar with a range of report formats, but they expect you to be self-sufficient and select the appropriate type. Your company will most likely have templates for report types. If the user's organization has established standards for the specific type of document you're creating, be sure to follow them.

Roy has a responsibility to be honest and accurate in his report. In spite of his desire to make the project look as successful as possible, he needs to keep his boss informed of the specifics, whether good or bad. Additionally, Carmen expects the status report to follow a standard format. This format differs from the proposal format because it contains up-to-date information on the project's progress. Roy also needs to review company policy to determine if any additional requirements have been established for status reports at Widget World. For his project to succeed, he needs to adhere to company standards for all project updates.

Other Types of Short Reports

Short reports fall into many different categories, but the common objective is to present information in a clear, accessible format. Some of the possible goals of a short report are as follows:

» To collect data
» To evaluate
» To provide a progress update
» To update sales numbers
» To announce an executive decision

The goal of the report plus the topic will determine the exact type of report. Three of the most common types of short report are status reports, research reports, and sales reports.

Status reports update a decision-maker on the progress of a project. Because of this, Roy needs to keep the summary section of his report current. He also needs to include a few areas of progress to highlight for Carmen. Often, status reports are required on some kind of recurring schedule, such as biweekly. The purpose of a status report is to give specific details that allow a decision-maker to follow the progress of a project, confirm that the project is in good shape, and anticipate or deal with problems.

Research reports present data about a specific topic. If an employee or a business has done a study about a particular topic, they can present their findings (typically to a boss or a customer) in the form of a short research report. This kind of report has similarities to a formal report, but it typically omits some sections, such as front matter and back matter. The goal of a short research report is to gather data into a simple format. Effective research reports are selective about what information to share without distorting the information. As the writer, you must understand which research findings are valuable to the audience. Roy highlighted key findings in his biweekly status report. Because he wants to keep the status reports concise, he saved the research details in case Carmen requests a full research report.

See Chapter **11** for more on front and back matter in formal reports.

Sales reports present findings about product sales. These reports help businesses maintain an accurate sense of their financial health. Typically, short sales reports are required on a regular schedule, but sometimes they are written under unique circumstances. For example, a company that launches a new product might track sales figures and want those figures in a simple format to share with its investors. The job of the technical communicator in this scenario is to present the sales numbers in an impactful (typically visual) way. The technical communicator should also anticipate and answer questions that might arise when the audience peruses the sales report (figure 10.11).

Beyond the three most common reports, figure 10.12 lists additional examples and how audience, purpose, and message vary for each.

Figure 10.11. Sales Report Spreadsheet with Graphs. The visuals in this sales report show data in a user-friendly format.

Sales Volume Report

Area	Volume	Last Year	This Year
Eastern	580	286	294
Southern	1119	436	683
Western	995	600	395
Northern	823	316	507
Middle	371	282	89
West-Northern	880	406	474
West-Southern	1063	586	477
East-Northern	1162	402	760
HK, Macow, Taipei	948	356	592
Oversea	602	584	18
Total	8543	4254	4289

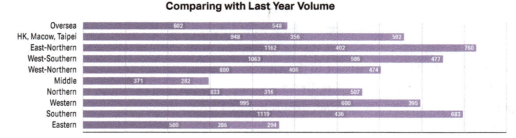

Figure 10.12. Types of Short Reports. This table lists additional types of short reports and how audience, purpose, and message differ for each.

Report Type	Topic/Message	Purpose	Audience
Audit	Business finances	To examine data	Owners, governing bodies
Feasibility report	Testing practicality of a project	To evaluate and recommend	Managers, governing bodies
Incident report	Formal explanation of an occurrence	To document	Owners, managers
Manager's memo	Formal statement of executive decision	To announce	Employees
Survey	Data taken via polling	To compile data	Researchers, managers

Typical Elements of a Short Report

The organization of short reports varies according to type. However, the following sections are common in most short reports.

Summary and Introduction

These opening sections explain why you wrote the report. The summary section, sometimes called the "purpose," offers an overview of the report's recommendations. The summary and introduction must be clear about its intent so the users understand how to interpret what follows.

Body

This section may contain a variety of components, including research findings, status updates, timetables, potential activities, or costs. The body of the report should share specific information rooted in data and analysis.

Conclusion

In the conclusion, the technical communicator reminds the user of the overall purpose of the report, reviews the key findings, and brings the report to a close. Some short reports conclude with a recommendation based on the findings. The recommendation should follow logically from the findings in the report's body.

Steps for Writing Proposals and Short Reports

See Chapter **7** for more on collaboration.

Whether you're writing a proposal or short report, the steps you take to create an effective technical document are much the same. The longer the document, the more likely the project will require collaboration. Be sure to plan additional time for coordinating with various members of your team.

Step 1: Plan

Determine a realistic scope, or purpose, before beginning. Narrow the focus of the topic to make it manageable. If you have a choice, opt for a specific rather than general topic. When planning, get input from your audience if possible. Talk to your audience directly, but if you can't, find other ways, such as reviewing previously successful projects and their reports, to determine what the audience wants to see in your document.

Step 2: Research

See Chapter **5** for additional research methods for technical communicators.

Locate convincing and relevant data. If your document gives the impression that you've only partially understood the topic or the audience, it will likely be dismissed. Just like when you write a college essay, outside sources help you make a strong case. What are the expert sources of information that you should consult? Who are the experts in the particular field? Are there authoritative sources of data that can be consulted? Research thoroughly and present your findings in a precise format that is cited according to your audience's expectation.

Step 3: Draft

Begin drafting the document once you have an adequate understanding of the topic and the needs of the audience. Ideally, the first draft should be written quickly. The goal in an initial draft is to get the information out of your head and onto the page. Avoid second-guessing yourself in an initial draft. Aim for a reasonable target word count. You might want to write a little more than you think you need in this stage because some sections will likely be cut during revision and editing.

Step 4: Revise

Revise the document by focusing on the big picture. What is the purpose of the document? Look at the draft and determine if your purpose is clear and direct. It's often valuable to get outside feedback at this point, especially from someone with experience in the topic. Review the needs of the audience. What do they expect to see in the document? What will convince them that your idea is valuable? What might be their concerns or interests? Look at the draft as a whole, and consider whether you have addressed these questions.

Step 5: Edit

Edit the document with a focus on precise paragraphs, sentences, and words. This step is where you take care of the details. Editing focuses on grammar, punctuation, style, and format at the sentence and word level. A lack of editing makes a poor first impression that could result in your document being rejected based on its presentation. In addition, consider whether your word choice is appropriate for the specific industry. Does your vocabulary reflect an appropriate level of familiarity with the topic and the audience?

Principles for Proposals and Short Reports

You can improve the quality of these documents by following a few basic principles, which include using precise language and an objective tone. With both of these principles, the emphasis lies in accuracy and fairness. The user must feel confident interpreting the meaning of a document. You can make this easier by examining your word choice and presentation of information.

Use Precise Language

The success of a proposal or short report hinges on specificity. Users need to know precise details that inform and lead them, when appropriate, to make a decision.

Be Specific

Avoid making sweeping statements based on insufficient evidence, otherwise known as generalizing. For example, don't say that something is a certainty if data suggests that it's only highly likely. Those are two different things.

Be Accurate

Words such as "always" or "never" can be useful, but they can also be inaccurate. Modify assertions when necessary with words such as "frequently," "seldom," or "almost." If the qualifier makes the statement more precise, use it.

Maintain an Objective Tone

Biased or slanted language can destroy credibility and cause a user to reject a document. Maintain objectivity to build a strong foundation for proposals and short reports. Even when you seek to convince the user in a proposal, your tone should remain objective.

Eliminate Opinion

Phrases such as "in my opinion" or "in our estimation" are overused in short reports and proposals. Instead, state information objectively. When persuading, use facts as a basis for opinions. Instead of stating, "we believe that moving to a virtual machine system is the best decision," use data to demonstrate that it's the best decision: "This graph shows that investing money now in a virtual machine system will save Widget World $50,000 in two years."

Include Experts

Use research to incorporate data and expert opinion to support your idea. Asking the user to trust the information in your proposal or short report based solely on your opinion is usually a mistake. Cite information from trusted sources to support your claims.

Present Information Ethically

Maintain your reputation as a reliable source of information by following the principles of accuracy and objectivity in your reports. When Roy gives credit to the creator of the image he uses in his report, he shows that he's paying attention to details and knows how to handle outside information properly.

Never present information from outside sources as your own. Cite sources accurately and consistently. This practice shows that you are a professional and thorough researcher.

Be Honest and Transparent

Don't cover up negative issues or problems. If a topic deals with a problematic situation, don't ignore the difficulties. Rather, present a realistic solution. This approach will increase your credibility in the eyes of the user. Also, it's likely that the user was already aware of the problem, so avoiding it would be a setback anyway. Transparency is crucial, so always communicate with honesty.

Case Study

Short Reports at Work

This case study is an opportunity for you to put into practice what you've learned. Part of this chapter focuses on creating short reports and examining the elements they share with proposals. Look at the following case study to consider how to include the expectations of your audience as you write a short report:

See Chapter **7** for more on workplace documents

At Widget World, Roy presents a finished proposal to his boss, Carmen. After receiving approval from the board of directors, Carmen tells Roy to begin the project and submit a status report after two weeks. A status report is a short document that provides an update about an ongoing situation and typically follows a proposal. It delivers necessary information to Roy's boss in a simple format that can be quickly digested.

Roy's top priorities are to move the project forward, stay within budget, and limit disruption to the workplace. Roy wants to write the status report simply and clearly to avoid giving Carmen more information than she wants. As the project progresses, Roy intends to follow the steps of planning, researching, drafting, revising, and editing all related documents. This includes the status report that's due in two weeks. What suggestions would you make to Roy about his upcoming report?

Discussion

» What type of workplace document (Gantt chart, memo, email, etc.) would be most appropriate to use for the status report?

» If Roy encounters a problem with the project, should he include the information? Why or why not?

» Roy is friends with Carmen. They worked together at a previous company. Is it okay if Roy sends Carmen a quick text about the project status?

Checklist for Short Reports

Type of Report

☐ What type of information did the user request from you? Have you considered how this will determine the type of report?

☐ Is your report addressed to a decision-maker (most likely the person who solicited it)?

☐ How often will the user need updates to the report?

Considerations

☐ What is the length requirement?

☐ Would the use of visuals help explain difficult or lengthy content?

☐ Did you cite borrowed information?

Output of Report

☐ Do you need a print and digital option?

☐ If you need both a print and digital version, how will the content and design shift?

☐ Did you budget for the type of output requested?

Conclusion

Success in all types of short reports, including proposals, depends on your ability to inform your audience or justify change. Make communication choices that support your purpose for the particular document type. Study the situation of your user and communicate accordingly. Craft a message that fits within the expected parameters for that type of document so the user can easily understand it.

When you effectively create proposals and short reports, you can save users money and demonstrate your expertise. In Roy's case, he can gain the approval of his boss, help his company, and show that he's motivated and competent at solving problems. You, too, can demonstrate your ability to solve problems by creating convincing proposals and short reports that make it easier for those in charge to act.

Chapter 11

Formal Reports

Abstract: Formal reports are multipage documents that collect and interpret data. A formal report presents complex information and suggests a solution or makes recommendations. This task frequently involves collaboration and requires that you present your research and recommendations ethically. The creation of a formal report is as much about the process of planning and research as it is about organizing and writing the final product. It may seem like a lot to juggle at first, but don't worry. You have been practicing the elements of effective communication throughout this book. Now you just need to pull it all together into a formal report. You've got this.

Looking Ahead

1. Why Formal Reports Matter
2. Types of Formal Reports
3. The Formal Report Process
4. Writing the Formal Report

Key Terms

» abstract
» analysis (causal, comparative, feasibility)
» appendix
» bibliography
» causation
» conclusion
» correlation
» end matter
» formal report
» front matter

» glossary
» index
» introduction
» letter of transmittal
» levels of evidence
» methodology
» objectivity
» research (primary/ secondary; qualitative/ quantitative)
» report body

» results
» scope
» subjectivity
» table of contents
» title page
» table of contents
» title page

Why Formal Reports Matter

Employers often rely on technical communicators to research complex issues and present their findings in an easy-to-digest format. The assignment often results in a formal report, sometimes called a long report or final report. Technical communicators are well-positioned to produce these reports because of their ability to study a subject and present ideas in a clear, concise, and accurate format.

Decision-makers, such as managers or business owners, can save time by asking a technical communicator to do research and present the findings. This assignment frees the decision-maker to do other important work. Even if you are not an expert on the topic of your report, the act of compiling the report can make you the resident expert in your workplace.

Formal Reports Defined

Formal reports have similarities to short reports, but formal reports are typically longer and take more time to develop. The purpose of a formal report is to gather and condense information relevant to a specific topic. Sometimes these reports also analyze information and make a recommendation. A typical formal report requires you, the technical communicator, to draw from all the skills you've encountered in this book.

Creating a formal report involves planning and researching, collaborating with and persuading others, and considering ethical situations related to the topic and your research. You must consider design, the use of multiple modes of communication, and how to acknowledge information from other sources. These reports require you to apply everything you've learned about technical communication, critical thinking, and research.

Here are a few examples from various fields where the formal report might be used:

> » An engineering firm needs to determine which buildings are structurally sound. They hire you to offer structural integrity solutions for buildings on a fault line.

» As a member of the safety committee at your school, you are tasked with examining the staffing of nurses within the school district. You must work with your committee members to review reports on student care, survey the nursing staff, and compile recent research from educational and medical reports.

» You are an analyst for your state's Health Authority, and you have been asked to compile a report on the opioid epidemic in your state. Your report makes recommendations for a new prescription drug monitoring program.

As you can see, formal reports are flexible documents used in many fields. Formal reports identify a problem or examine alternatives and then present the findings at length as a written solution or recommendation.

Formal Reports and the Problem-Solution Writing

The Problem-Solution Framework is fully realized in the formal report (figure 11.1). In the case of a formal report, the problem typically originates with a lack of information that prevents a decision-maker from moving forward. By considering audience and purpose, a technical communicator can create a message in the form of a document. This document should provide a solution to the user's problem.

Figure 11.1. Problem-Solution Framework. The formal report represents the quintessential example of the Problem-Solution Framework.

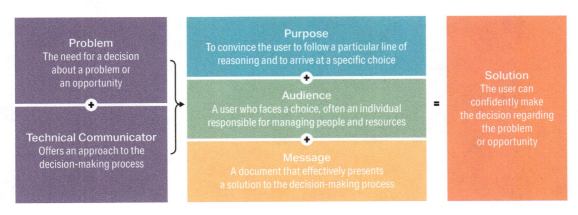

Problem
The need for a decision about a problem or an opportunity

+

Technical Communicator
Offers an approach to the decision-making process

→

Purpose
To convince the user to follow a particular line of reasoning and to arrive at a specific choice

+

Audience
A user who faces a choice, often an individual responsible for managing people and resources

+

Message
A document that effectively presents a solution to the decision-making process

=

Solution
The user can confidently make the decision regarding the problem or opportunity

Formal Reports at Work

In Chapter 5 you met Jessamyn, the technical communicator researching electric vehicles for Tomorrow's Taxi Company. In this chapter, she dives into a report that goes a little deeper. At her boss's request, Jessamyn will create a report that explores whether the use of autonomous vehicles (AVs) is a viable option for their company.

Jessamyn's boss wants to know if adding driverless cars, also referred to as AV, to the company's fleet is a smart move. Notice how the problem (a lack of information) leads to the purpose of the report (to find information). Now Jessamyn needs to identify the report's message, which is what her boss (the audience) wants to know. She could approach this analysis in a variety of ways. Her formal report could be one or a combination of the following forms: a comparative analysis, a causal analysis, or a feasibility analysis. The next section considers each of these at length.

Types of Formal Reports

Formal reports take different forms depending on their purpose, but all require some kind of analysis. **Analysis** means to look at how the individual parts of a complex process or product work together. For example, chemists regularly perform analysis on substances to identify their makeup on a molecular level.

As a technical communicator, your analysis will focus on how complex ideas or situations are made up of more specific components. Most technical analysis includes a certain amount of informed speculation about causes and future possibilities as well. Formal reports almost always include analysis, which is why the three varieties explored here have "analysis" in their name. Refer to this table for a quick overview of common analytical reports (figure 11.2).

Figure 11.2. Types of Analysis. Use this table as a quick reference for the three main types of analytical reports described in this chapter.

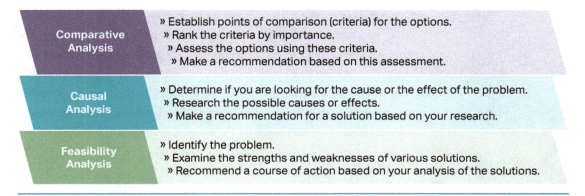

Comparative Analysis	» Establish points of comparison (criteria) for the options. » Rank the criteria by importance. » Assess the options using these criteria. » Make a recommendation based on this assessment.
Causal Analysis	» Determine if you are looking for the cause or the effect of the problem. » Research the possible causes or effects. » Make a recommendation for a solution based on your research.
Feasibility Analysis	» Identify the problem. » Examine the strengths and weaknesses of various solutions. » Recommend a course of action based on your analysis of the solutions.

Comparative Analysis

A **comparative analysis** is a formal report that establishes a set of criteria to look at similar items or situations to determine the best choice. Businesses often need a comparative analysis when considering a large purchase. For example, if a business wishes to expand to an overseas market but can't decide between two specific locations, it might request a comparative analysis. Or consider an organization that wants better healthcare coverage for its employees. In this case, a comparative analysis could weigh the options and help the organization make the best choice.

If Jessamyn conducts a comparative analysis, she will need to determine how to compare the cars in the company's current fleet to potential AV. Jessamyn must establish points of comparison up-front so that the different vehicle types are compared fairly and consistently. She will likely compare the cost of purchase or replacement, fuel, insurance, maintenance, and labor. She'll determine how much revenue each vehicle generates (or is estimated to generate) and its reliability. Then, in the report, she will clearly explain the criteria she used in the methodology section. In her report, she will rank the cars according to the criteria, assess their strengths and weaknesses, and end with a recommendation of the best option.

A comparative analysis weighs the evidence between two or more ideas, situations, or products and makes a claim based on evidence about which one is best. What's best for one group may not be best for another, so it's important to define the criteria and audience at the beginning. For example, if Jessamyn's research points to AVs producing significant cost savings, the decision to go with AVs may be best for the company. However, it may not be the best for the drivers who would be replaced. In order for a comparative analysis to be effective and ethical, the items or issues under comparison must be measured by the same standards.

Causal Analysis

A **causal analysis** looks at why something happens or could happen. Often this variety of formal report analyzes effects or what might occur if a particular decision is made. In other cases, the causal analysis considers what already took place to help decision-makers understand how to prevent future problems or take advantage of opportunities.

For example, if important machinery breaks down in an industrial business, a causal analysis is often conducted to figure out what happened. The report helps the business determine what to do in the future to protect the machinery and the people who operate it. Such a report might also allow the business to plan for better maintenance or replacement of the machinery.

In Jessamyn's case, a causal analysis might investigate what is causing the current fleet to break down as a part of measuring reliability and revenue. She might also look at ethical issues of AVs or liability issues with insurance companies and how these issues impact the company's bottom line. Jessamyn explains the research in the findings section and analyzes the research in the report's discussion section, which helps the decision-maker know if they want to move forward.

Determining cause requires careful analysis. There may be direct and indirect causes. There may be one cause or multiple contributing factors. The researcher must consider what is relevant and avoid rushing to judgment. Causality must be clearly demonstrated through cause and effect evidence.

Beware of confusing **correlation** (two things connected by circumstances) and actual **causation** (one thing directly impacts something else).

Feasibility Analysis

A **feasibility analysis** determines if a strategy, plan, or design is possible. Is the proposed decision a good idea for a business or client? Is it economically justifiable? Will the strategy, plan, or design produce the desired results? A feasibility analysis can be invaluable for a business that is weighing the possible benefit of a risky decision. In the example of the organization considering the best option for providing healthcare for its employees, a feasibility analysis might determine whether the organization can afford to go with the best provider and package.

Jessamyn's formal report that serves as the example for this chapter is primarily a feasibility analysis. Her job is to determine if AVs are a financial, marketable, and operational solution to the fleet's problems. To make a solid recommendation, she will have to look closely and honestly at both the strengths and weaknesses of using autonomous vehicles.

The technical communicator must objectively assess the possible benefits or drawbacks of a particular course of action, without allowing their personal opinion to cloud the final conclusions. On a personal level, Jessamyn finds the idea of driverless vehicles unnerving, but she needs to set aside her opinion as she collects and presents her data. A feasibility analysis also considers alternate points of view in the decision-making process. The best feasibility studies will weigh the pros and cons of the collected data before recommending a course of action. Beware the ways bias can sneak in when you interpret data or draw conclusions. Analysis requires a steadfast commitment to objectivity.

The Formal Report Process

This section outlines steps to create a report. Familiarize yourself with the different sections so you can be effective at each step along the way.

Develop a Plan

To develop a plan, ask yourself three questions:

» How much time do I have?
» Whom do I need to talk to?
» What information do I need first?

Before you dive into the deep work of your report, give yourself time to plan. Consider how much time you have and what you need to accomplish. If you don't know, confirm the deadline for the report with your manager.

Often, a formal report will involve multiple contributors. Generate a contact list for others involved in the project, and if necessary, determine what part of the report they will be completing. Formulate a basic outline of how you'll complete the project, even if it may change later. The rest of this section offers more steps that can help you develop a plan that you can customize to the project.

Jessamyn's outline determines what actions to take and in what order. She makes a list of everyone she needs to talk to and calculates how much time it will take to gather and analyze the information from various departments. She builds in a little extra time to her total—20 percent should do it—because she knows that collaborative work takes longer. With this in mind, Jessamyn sets reasonable deadlines for herself and others.

Like Jessamyn, if you want to stay on target, you need to begin with a question. Sometimes your boss determines the question when requesting the report. Other times, you determine the best course of action, beginning with a question of inquiry. Take a moment to consider the main question that you are trying to answer. This typically leads to other smaller questions, which should be addressed in the analytical section of your report.

Jessamyn's main question is whether it is feasible to add AVs to the fleet. As she searches for an answer, many other questions arise:

» How much would a fleet of AVs cost?
» How would they operate? Are they safe?
» Would they save the company money on repairs, insurance, or staffing?
» What type of fuel and maintenance would they require?

» Would the community, their customer base, support the idea?

» Would they be likely to increase company revenue?

The challenge is to determine which questions directly relate to the issue. Jessamyn needs to prioritize her questions. One way to do this is to carefully consider the report's purpose.

Determine Your Purpose and Scope

To determine your purpose and scope, ask yourself the following questions:

» Where should my research focus?

» What is my report intended to do?

» What are the boundaries for my research?

Asking *why* will lead you to the **purpose** of the report. You can think of the purpose as similar to a thesis in your academic essays. Your purpose explains why the report exists and what it's about. Establishing a clear purpose early in the project will help you stay focused and communicate clearly. Often, reports will have a primary purpose and a secondary purpose. For example, if the primary purpose of a comparative analysis is to consider two different building designs for a future business location, a secondary purpose might be to introduce a third design option.

In Jessamyn's case, her boss wants to know if AVs are a better option for Tomorrow's Taxi Company. The report's primary purpose is to determine whether making a change will lead to greater profits. The report's secondary purpose might be to identify other factors to produce greater profits. Make sure the solution you propose aligns with the purpose of the report. Jessamyn doesn't want to suggest a solution only about the advantages of vehicle features if the primary purpose of the report is about the company's profits.

Scope refers to the type and amount of information included in your report. At some point, you will need to decide what belongs in the report and what does not. You can arrive at your project's scope by asking what type of information the end user needs and does not need. Too little information leads to an uninformed decision. Too much information can be overwhelming, unnecessary, and a waste of time.

Identify Your Audience

To determine your audience, ask yourself these questions:

» Who is the primary user of this report?
» Is there a secondary audience?
» Is there anyone else who needs to know about my research?

See Chapter **①** for more on creating a user profile.

Your audience determines the tone and language of the report. Use language that is understandable to the majority of your users. If you are using technical terms that are widely known, you need to define them. Knowing as much as you can about your audience can also help you determine the most effective tone and word choice. Usually, this means using neutral, unbiased language that is accessible based on your audience's level of technical expertise.

Jessamyn's boss or a committee may be the primary audience, but the document might also be read by other departments, such as HR or finance. Rarely will someone read a report from start to finish. Your audience is likely to read certain sections that are relevant to them, depending on their relationship to the problem and their decision-making responsibilities within the organization.

Most of Jessamyn's audience knows little about autonomous cars, though many know about traditional ones. Most terms specific to the auto industry will be easily understood while those specific to the new technology will require explanation.

State the Issue

To identify the issue, ask yourself these questions:

» What is at stake?
» What problem am I trying to solve?
» Can I explain the issue in a single sentence?

A statement of the issue can keep you focused during your project. For your formal report, create a short, specific issue statement as a guide. This statement should result from the steps that came before, so form it after considering your purpose, scope, and audience. For example, if a technical communicator completes a causal analysis regarding a business's loss of productivity, they might state the issue as follows: "This report analyzes the factors that led to a 15 percent loss of employee productivity during the past fiscal year." The solution may advise this number is reversible by providing additional training for supervisors.

Jessamyn's purpose is to determine the feasibility of autonomous cars. Why? Because the current fleet is becoming expensive, contributing to air pollution, and reducing the company's revenue. As Jessamyn has done here, aim to keep your statement of the issue direct and short. It will help you stay on track as you begin your research.

Conduct Research

To begin conducting research, ask yourself these questions:

» What sources will reliably give me the information I need?
» Do I have access to an expert to interview?
» Have I included a variety of sources?

The type of research required for a formal report depends on the type of report and your field or profession. Generally, research falls into two categories: primary research and secondary research. **Primary research** is the type that you collect yourself directly from a source, such as through interviews, surveys, or observations. **Secondary research** is the use and synthesis of previously collected data—other people's surveys, studies, opinions, or observations.

Whenever possible, combine primary and secondary research. Secondary research provides a background understanding of your topic. Primary sources add to what is already known and build your credibility. This combination of research allows you to explore the topic from a variety of angles.

Jessamyn thinks her company needs to conduct surveys in the community to see if their customer base will support the idea of AVs. This is an example of primary research. A survey would help add validity to Jessamyn's recommendation by bringing in a local voice. Ideally, she examines surveys done by other credible sources and considers them along with the data she collects on her own.

To accompany the community survey responses, Jessamyn needs to gather and organize specific data on the current fleet, so she interviews a program analyst at the Department of Motor Vehicles (primary research). She also gathers financial reports and estimates from other departments, as well as current accident statistics and studies on driverless vehicles, as these are readily available from reliable sources (secondary research).

Quantitative and Qualitative Research

Research is generally organized into two main categories that appeal to different parts of our brains. Most reports require a combination of quantitative and qualitative research. **Quantitative research** aims to measure or prove something with numerical evidence. Think quantity. Numbers. **Qualitative research** aims to describe or give depth to a topic through data that does not include numbers. Think quality. Observable fact. Depending on the issue and the recommendation, one form of research might be better suited for your report. Or you might choose to include both types of research (figure 11.3).

Figure 11.3. Types of Research. Most reports require a combination of quantitative and qualitative research. Quantitative research aims to measure or prove something. Qualitative research aims to describe or give depth to a topic.

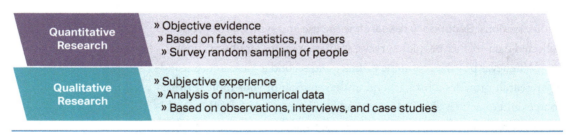

Jessamyn knows that much of her research will be quantitative and focus on facts and numbers, such as direct cost and accident statistics. Yet there are qualitative elements that are essential to consider. Jessamyn consider qualitative elements like customer satisfaction, expert opinions, and ethical issues of AVs, all of which could impact revenue.

Objectivity and Subjectivity

Researchers need to practice objectivity and recognize when they, or the sources they consult, have moved into subjective interpretation. **Objectivity** means to arrive at an understanding of a topic based on external evidence or verifiable facts. **Subjectivity**, on the other hand, means to consider a topic based on personal perceptions and interpretations.

In Jessamyn's case, an objective statement would be "AVs reduce air pollution," which she can back up with fact. A subjective statement would be, "AVs will scare away customers." While that might be a reasonable conclusion, there's no data on that yet. It's easy to cross over from objectivity to subjectivity without realizing it, which is why it is so important to analyze the information you find.

Analyze the Information

To analyze the data, ask yourself: What is this information telling me? What does it mean?

Here are some additional questions to help you consider the information you've gathered:

» Is it thorough enough and consistent with the criteria I've established?
» Does it help answer the focus question?
» Is it enough to draw a conclusion and make recommendations?
» Have I presented the information ethically and without bias?

Once you have considered these questions, you can also consider the levels of evidence (figure 11.4). **Levels of evidence** (sometimes called a hierarchy of evidence) are assigned to studies based on the quality of their methodology. In other words, studies are evaluated by how rigorous and precise their research methods are. These categories are used in evidence-based medicine, but it can also help you analyze data you're collecting for any kind of evidence-based report.

Figure 11.4. Levels of Evidence. Peer-reviewed academic journals are the highest level of evidence because they are produced by scholars in the field and reviewed and evaluated by other scholars prior to publication.

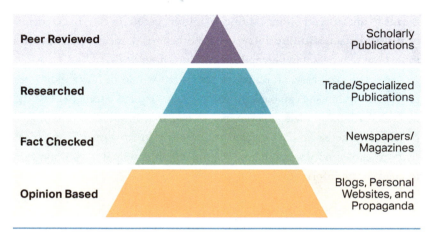

The most reliable evidence includes scholarly sources that are peer reviewed. Trade journals and even government documents are generally produced by experts in the field, but they don't go through the same rigorous review process. Popular sources like magazines and newspapers (print or online) are not as reliable because they tend to be more opinion based.

When considering what information is essential to your report, remember your scope. Information that merely adds to the length or variety of the report is unnecessary. After spending so much time collecting data for this report, Jessamyn struggles to decide what is relevant and what is not. This struggle is not a sign of failure. At this stage, Jessamyn may need to take a draft of the report to a trusted colleague who can look at it with fresh eyes and provide feedback.

Draw Ethical Conclusions

To draw ethical conclusions, ask yourself these questions:

» Have I remained objective?
» Does the research drive my conclusions?
» Are claims that I made supported with quality evidence?

After analyzing your data, it is time to consider what the information tells you about the strengths and weaknesses of the idea or the cause of the concern. Let your research drive the conclusions. It's important to look at the information without bias and to be fair and honest in the treatment of the material, even if it does not align with your preferences or your boss's preferences.

Despite her initial hesitation, Jessamyn recognizes the value of AVs and supports the idea of using them to replace the current fleet. However, the evidence doesn't demonstrate their clear advantage. The technology is new and expensive, and the regulations regarding the vehicles are fluid. Essentially, the vehicles are still in the pilot phase. Her job is not to talk her boss into her preference or stack the evidence in favor of it. Instead, she must view and present the information rationally and ethically for the best interest of the company.

Decide on Recommendations

To make a recommendation for a solution, ask yourself these questions:

> » Did I provide enough information for decision-making?
> » Is the recommendation consistent with the issue?
> » What is the best option based on the research?

Most varieties of formal reports include a recommendation. Base your suggested course of action on what you've concluded from your analysis. Again, remember your purpose. Form your recommendation on the best option or most reasonable cause. Sometimes you will not be able to make a recommendation due to limited information. Sometimes the study does not yield a definite solution. Your recommendation might be conditional or tentative, depending on other factors.

For your recommendations to be convincing, you need to show the connection between your findings and your suggested actions. Don't make recommendations based on gut feelings or impressions. Instead, show how your suggestions directly result from your analysis of the data. The users who study your document need you to convince them that your recommendations are reasonable, so highlight the connection between data and recommendations. Be overt, and don't assume that the connection is clear. Often, a graphic representation of the data from your research can underscore your recommendations.

Based on her research, Jessamyn knows now is not the right time to add high-tech AVs to the fleet, but she sees the potential in a few years' time. The information suggests that the industry will eventually head in this direction and that the company will need to be ready when the time comes. She decides to recommend that the company begin a slow transition process, but only if the technology continues to improve.

Writing the Formal Report

As you write the report, ask yourself these questions:

- » Do I have all the parts I need?
- » Do I have the parts in an expected order?
- » Have I provided memorable information?

The report consists of front matter, the body, and end matter. The following example provides an outline that can be adapted for just about any formal report (figure 11.5). Even though the sections may have different names depending on the type of report or the field you're working in, they serve similar purposes.

Front Matter

Front matter introduces the report's contents. It also guides users to relevant information in the report. Most users will be looking for specific information in the report, and the front matter helps them find it easily. Front matter generally includes the following components:

- » The **letter of transmittal** is written to the user or users who requested the information (see figure 11.6). It's like a cover letter for your report.
- » The **title page** includes the title of the report, the authors and their organization or department, the person or organization for whom it was prepared, and the date.

» The **table of contents** lists the sections of the report and their page numbers. Your headings should be orderly, consistent, and clear.

» The **abstract**, or executive summary, is a brief summary of the report that includes the problem, method of analysis, results, and conclusion or recommendation.

Figure 11.5. Formal Report Outline. This outline shows the most common sections in formal reports. This chapter will explore each in more detail.

Outline for Formal Report

1. Front Matter
    a. Letter of transmittal or cover letter
    b. Title page or cover page
    c. Table of contents
    d. Abstract orexecutive summary

2. Body
    a. Introduction
    b. Background and definitions
    c. Purpose and target audience
    d. Methodology
    e. Scope
    f. Findings
    g. Summary/discussion
    h. Recommendations/solutions
    i. Conclusion

3. End Matter
    a. Works Cited, References, or Bibliography
    b. Appendix
    c. Glossary
    d. Index

Your report may have more than one Findings and Discussion section.

Figure 11.6. Letter of Transmittal Template. This model provides guidelines for a standard letter of transmittal. This letter introduces your formal report and is addressed to the person who requested the report.

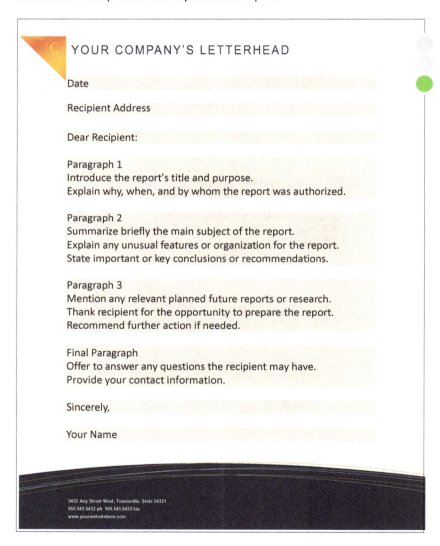

YOUR COMPANY'S LETTERHEAD

Date

Recipient Address

Dear Recipient:

Paragraph 1
Introduce the report's title and purpose.
Explain why, when, and by whom the report was authorized.

Paragraph 2
Summarize briefly the main subject of the report.
Explain any unusual features or organization for the report.
State important or key conclusions or recommendations.

Paragraph 3
Mention any relevant planned future reports or research.
Thank recipient for the opportunity to prepare the report.
Recommend further action if needed.

Final Paragraph
Offer to answer any questions the recipient may have.
Provide your contact information.

Sincerely,

Your Name

5432 Any Street West, Townsville, State 54321
555.543.5432 ph 555.543.5433 fax
www.yourwebsitehere.com

Jessamyn has a draft of her front matter, but her table of contents needs a little work. Take a look at the model that follows to see if you notice what needs improvement before it goes to her boss (figure 11.7).

Figure 11.7. Draft Table of Contents. To avoid errors on your table of contents, use the automatic table of contents tool in Word. The mistakes you see here are a result of manually typing the table of contents.

TABLE OF CONTENTS

Executive Summary..............................iii
Introduction...1
Methodology......................................1
Limitations......................................1
Background...1
What is an Autonomous
Vehicle?..2
Findings...3
Cost of Autonomous Vehicles..........3
Insurance Rates.........................3
Maintenance.....................................3
Safety..4
Safety Concerns and Controversies.......4
Market Analysis.................................4
Regulations...5
Cost Comparison................................6
Discussion of Known and Unknown Cost Factors
Affecting the Comparison..........7
Conclusion...7
Recommendations.................................8
Option One – Partnership and Pilot........8
Option Two: Test Vehicles..............8
Option #3 – Partial Replacement......8
Works Cited....................................10
Appendix A...12
Table 1: Estimated Revenue for an Autonomous Vehicle
Fleet...6
Figure 1: Society of Automotive Engineers Automation
Levels...12

The page numbers are not properly aligned.

The three recommendation options use faulty parallelism. The options are formatted and punctuated differently from each other and are not italicized like others of the same heading level.

The bold text makes the table and figure look like sections of the report. The font and spacing of the table of contents is large and forces the list of tables and figures to be too close to the other material.

In addition to the annotations on the model, you may have noticed some consistency issues in the headings. The subheading "Safety Concerns and Controversies" is under the "Safety" heading, so the repetition of "safety" is unnecessary. Jessamyn should aim to keep her headings as concise as possible. Turn to the end of the chapter to see how she revises her table of contents.

Body of Report

The report consists of an **introduction** that describes the problem and defines the purpose, scope, background for the topic, and methods of analysis. If terms require definition, they will be included in the introduction. If there are five or more terms, these will be included in a glossary instead, which is located in the end matter.

Methodology explains how you gathered data, including what tools you used, what sources you consulted, and what research strategies you defined. This section is where you explain how you went about assembling your report.

Establishing your method does two things. First, it helps you control the scope of the analysis. Second, it helps the user to understand the process used to arrive at conclusions in the report. This is important because users often want to consider the value of data and analysis in light of the methodology.

You can see how Jessamyn begins her report with a draft of the methodology section (figure 11.8). Note how the design makes it easy to scan the areas to be evaluated, as well as the limitations for the report. Jessamyn's introductory statement helps the user understand how she is connecting the purpose of the report to the specific areas of evaluation. This is her first draft of the section, though. She still needs to polish it up a little more. Turn to the end of the chapter to see how she revises her entire report.

Figure 11.8. Draft Methodology Section. Jessamyn has determined criteria for her methodology, but she needs to shorten her bulleted list for readability. Her inclusion of limitations, however, is a good move. It shows her awareness of the complexity of the topic.

Methodology

To weigh actual costs vs. benefit(s) of this option, the following areas must be considered:

- The annual cost of the current fleet in routine maintenance, repairs, insurance, and fuel
- The annual cost of labor to drive and maintain the current fleet, as well as train new drivers
- Estimated opportunity losses (lost revenue) for downed vehicles
- The initial cost of replacing the current vehicles with autonomous ones
- The estimated annual cost of the autonomous fleet in routine maintenance, repairs, insurance, fuel/energy, and operation
- The safety risks and benefits to clients
- Market reaction
- Potential regulatory changes or restrictions within the city and state
- The estimated gross revenues of the autonomous fleet vs. the known revenues of the current fleet

Limitations

This technology is new and while several companies are working to produce these vehicles, many are still in the development phase. There are few vendors for driverless vehicles at this time, and the estimated costs to purchase, insure, and maintain them are estimated or theoretical.

> While the bullets do break up the material visually, nine bullets are a bit much for the reader to absorb. Jessamyn should consider limiting her list to seven items or fewer.

> Parallelism in lists is another way to increase readability. Jessamyn should revise to make sure every bullet begins with a similar part of speech.

The introduction is followed by the **report body**, which includes the collected data, analysis of the findings, or results. This section is followed by a conclusion.

The **conclusion** ties the results together and can continue at length if it includes steps for implementation or a proposed plan. When requested or appropriate, recommendations are made as part of the conclusion. A **results** section includes background on the issue, data, expert opinions, and an interpretation of the material. Essentially, this section presents what you learned that will help the end user make a decision. The specifics of the results will vary depending on the type of report.

Jessamyn's report ends with recommendations (figure 11.9). She describes three possible options and then concludes with an evaluation of these options.

Figure 11.9. Draft Recommendations Section. The recommendations section of the formal report is likely to get a lot of attention. Be sure to provide a clear statement about what should be done.

conservative and will yield the lowest revenue while allowing the community time to adjust and time to retrain or reallocate drivers. The third enables the company to keep 100% of its profits, but with less oversight and assistance from the manufacturer, it encompasses higher risk. All three provide the opportunity to test the waters of the industry before fully committing financially.

Recommendations

The following options are the most feasible at this time:

Option One – Partnership and Pilot

"Corporation X" seems to be the U.S. automotive company most eager to release a fleet of AVs. They've developed partnerships with a few other companies in order to pilot their vehicles and are interested in using Tomorrow's Taxi Company as well. For the first year, they would provide the thirty-eight vehicles. Tomorrow's Taxi Company would be responsible for maintenance and energy costs, insurance, as well as monitoring staff. They would oversee any issues with the vehicles, make changes and upgrades, provide parts, and train our staff. They would receive .5% of fare profits for this period for the use of the vehicles. At the end of the contract period, it could be renegotiated for another year if the technology is still needing improvement (which would likely be the case) or Tomorrow's Taxi Company would have the option to negotiate a purchase for the vehicles at a discounted rate.

Option Two – Test Vehicles

Under this scenario, Tomorrow's Taxi Company would lease-to-own two driverless vehicles from "Corporation A" at a rate of $20,000 per month for a period of six months. For the first six months, the corporation would handle the monitoring of the vehicles remotely. After the trial period, the company would have the option of extending for another six months or paying the balance on the vehicles and purchasing the monitoring equipment. If leased for an additional six months, then an end-of-term buyout of return of vehicles would be required.

Option Three – Partial Replacement

In this option, Tomorrow's Taxi Company would replace five of its vehicles that have the highest mileage with AVs and purchase the monitoring equipment at an estimated cost of $1,200,000. This would include replacing the two minivans currently used in the hilltop community, as these are experiencing significant wear and tear on brakes. The company would not reap the same benefits as replacing the whole fleet, but would have the opportunity to do a trial run. The first option is the most aggressive. Though the vehicles would not be purchased initially, it does require commitment to the concept. Once the drivers and current vehicles have been reallocated to another division or become unavailable, it will be difficult to return to previous operational systems. The second is the most

Not all feasibility reports will require multiple options.

These paragraphs are getting a little too long. Jessamyn should break these up to make the text more readable.

End Matter

End matter includes anything after the main body of the report. Examples of end matter include some or all of the following items:

» A **bibliography** is a list of the sources used for research in the report. This could be a list of "References," if citing in APA, or "Works Cited," if using MLA. There are many other documentation styles that are industry specific. The title and format will depend upon the citation style.

» An **appendix**, or appendixes if you have more than one, provides information that offers further explanation or reference. Appendixes may include maps, complicated formulas, specific questions and answers in a survey, or reports that were referenced in the body. These extras may be of interest to specific users who want to better understand how you came to your conclusion or to verify your methods. When writing about information that appears in the appendix, be sure to provide a parenthetical reference, like so: (see Appendix A). Be sure each appendix is given a title and, if there is more than one appendix, a letter.

» A **glossary**, an alphabetized list of specific terms with definitions, can be included if your report has more than five terms that need to be defined. Otherwise, the terms can be defined in the introduction.

» An **index** is a list of specific items or terms within the body and the page numbers where these items or terms appear. An index enables users to search the report by topic.

Revised Professional Formal Report

Look at Jessamyn's revisions in her final formal report (figures 11.10a through 11.10s). The report in this chapter uses APA style for formatting, in-text citations, and the References page. You can find the same report using MLA style in the appendix. Your instructor may have other requirements for format and style—always follow the instructions provided by your course assignment.

Figure 11.10a. Revised Letter of Transmittal. This is Jessamyn's revised letter of transmittal.

November 15, 2020

Morgan Milford, CEO
Tomorrow's Taxi Company
1111 NE Rogers Dr.
Averagton, AZ 62000

Dear Mr. Milford:

Enclosed is my report, "Feasibility of the Implementation of Driverless Vehicles," for your review. Thank you for giving me this opportunity and allowing me to further my knowledge and the knowledge of our organization on this very important issue.

As you will see, some of this report's information is theoretical because driverless vehicles are not yet commonplace and are operated primarily by large automotive technology companies. Taxi companies are just beginning to join these ranks, and I'm confident that the material enclosed will help Tomorrow's Taxi Company decide when to make a similar leap.

I would like to thank our marketing, service, payroll, and human resources departments for contributing to this report. Their knowledge and data are essential to our decision.

If you need any additional information, please contact me. I can be reached at sanchez.j@ttc.com or 555-222-1000, ext. 200. Thank you for your time and consideration in reviewing this proposal.

Sincerely,

J. Sanchez

Jessamyn Sanchez
Director of Business Development

Get straight down to business with the first sentence of your letter of transmittal.

Show that you are a good colleague by naming the individuals and departments that contributed to your report.

A letter of transmittal should not be longer than one page.

Figure 11.10b. Professional Title Page in APA Style. This is Jessamyn's title page for her formal report. APA has a different format for professional and student title pages. See the inset for a student title page template. Check with your instructor to see which title page format they prefer.

Use a shortened version of the report's title for the running head for professional reports. The title page is considered the first page of the report.

The report's title should be in bold and centered. The first, last, and all other important words should be capitalized.

The title page should include an author note in bold that indicates who prepared the report and provides contact information.

Student title pages do not include a running head or author note.

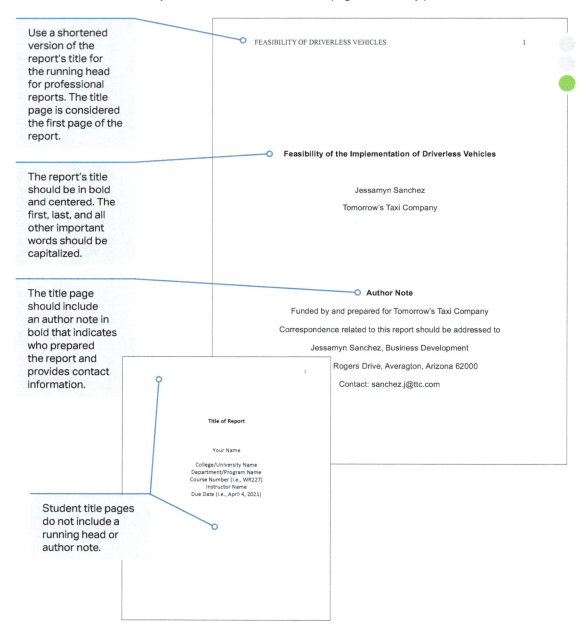

FEASIBILITY OF DRIVERLESS VEHICLES   1

Feasibility of the Implementation of Driverless Vehicles

Jessamyn Sanchez

Tomorrow's Taxi Company

Author Note

Funded by and prepared for Tomorrow's Taxi Company

Correspondence related to this report should be addressed to

Jessamyn Sanchez, Business Development

Rogers Drive, Averagton, Arizona 62000

Contact: sanchez.j@ttc.com

1

Title of Report

Your Name

College/University Name
Department/Program Name
Course Number (i.e., WR227)
Instructor Name
Due Date (i.e., April 4, 2021)

Figure 11.10c. Revised Table of Contents. This is Jessamyn's table of contents for her formal report.

FEASIBILITY OF DRIVERLESS VEHICLES 2

Table of Contents

Abstract 3
Introduction 4
Methodology 5
    Limitations 5
Background 6
    What is an Autonomous Vehicle 7
Findings 8
    Cost of Autonomous Vehicles 8
    Maintenance 8
    Safety 9
      Concerns and Controversies 9
    Market Analysis 10
      Regulations 11
      Cost Comparison 12
Discussion 14
Recommendations 16
    Option One – Partnership and Pilot 16
    Option Two – Test Vehicles 17
    Option Three – Partial Replacement 17
Conclusion 18
References 19
Figures 21

The table of contents lists the headings and subheadings as they appear in the report.

This report includes three levels of headings.

This report is written in APA style, so it includes a page titled References.

Figure 11.10d. Revised Professional Formal Report in APA Style. This is Jessamyn's abstract, also called an executive summary.

Most abstracts are about 150 words and are written as a single paragraph.

Do not indent the first paragraph of the abstract.

In a professional report, keywords are provided to describe the important ideas in the report.

FEASIBILITY OF DRIVERLESS VEHICLES   3

Abstract

Due to the age and high cost of maintaining its current fleet, Tomorrow's Taxi Company looks to determine the feasibility of replacing their current vehicles with autonomous vehicles (AVs). The technology is new and still developing, but recent advances mean the technology is becoming safer for customers and more cost effective to operate. However, AVs are not yet readily available and affordable. The primary players in this field are large automotive and technology companies with large amounts of capital and industry support, though they are beginning to offer their vehicles to other companies, such as taxi and delivery services. Clearly, the market is moving quickly in this direction. Costs, which are quite high at this time, are expected to become more competitive within the next few years. It is recommended that the company begin cautiously transitioning into the driverless market and preparing its business model and employees for this change.

Keywords: automation, automotive industry, autonomous vehicles, driverless technology, economic assessment, greenhouse gas emissions, state and federal regulations, traffic safety

Figure 11.10e. Revised Professional Formal Report in APA Style. This is Jessamyn's introduction for her formal report. You generally do not find cited material in the introduction. It is your opportunity to introduce relevant information in your own words.

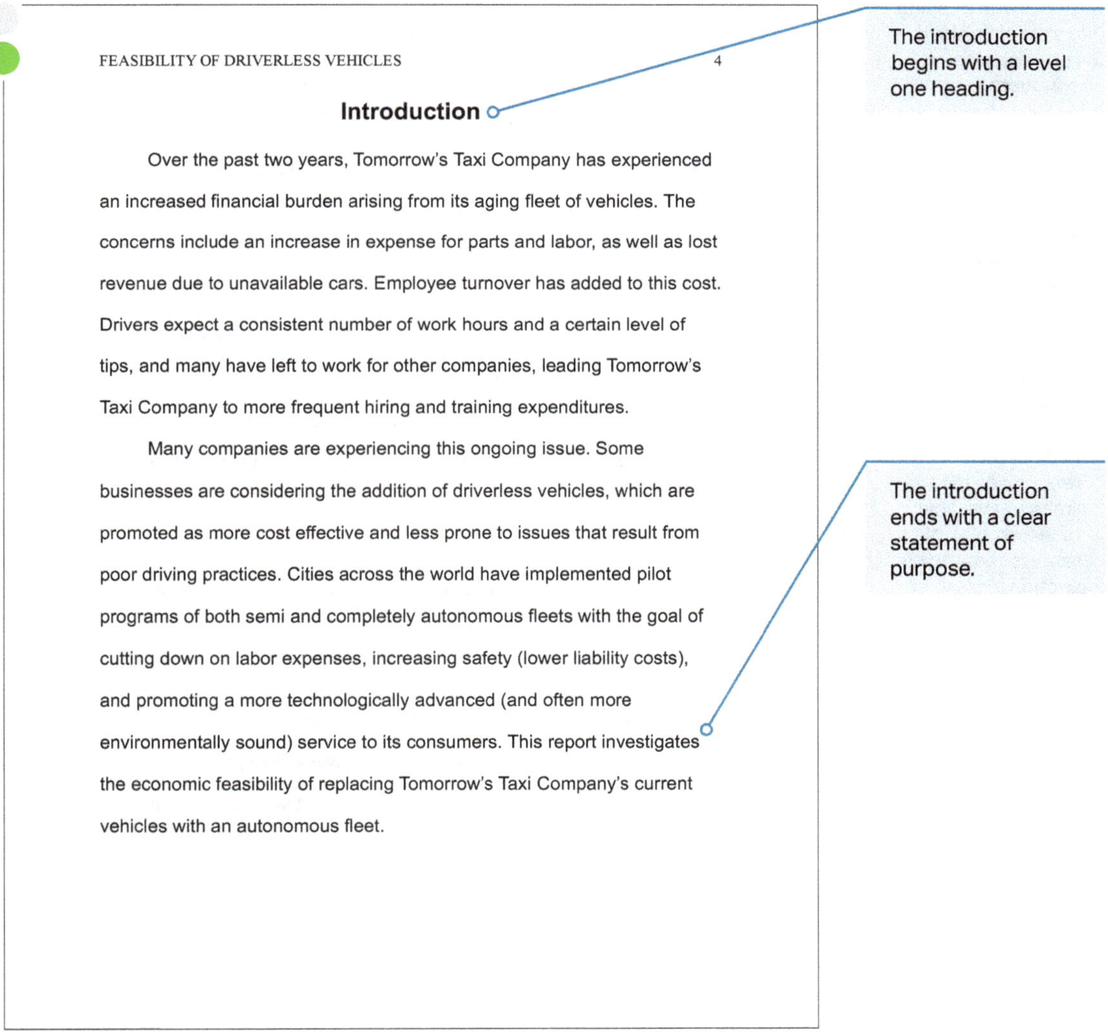

FEASIBILITY OF DRIVERLESS VEHICLES   4

Introduction

Over the past two years, Tomorrow's Taxi Company has experienced an increased financial burden arising from its aging fleet of vehicles. The concerns include an increase in expense for parts and labor, as well as lost revenue due to unavailable cars. Employee turnover has added to this cost. Drivers expect a consistent number of work hours and a certain level of tips, and many have left to work for other companies, leading Tomorrow's Taxi Company to more frequent hiring and training expenditures.

Many companies are experiencing this ongoing issue. Some businesses are considering the addition of driverless vehicles, which are promoted as more cost effective and less prone to issues that result from poor driving practices. Cities across the world have implemented pilot programs of both semi and completely autonomous fleets with the goal of cutting down on labor expenses, increasing safety (lower liability costs), and promoting a more technologically advanced (and often more environmentally sound) service to its consumers. This report investigates the economic feasibility of replacing Tomorrow's Taxi Company's current vehicles with an autonomous fleet.

The introduction begins with a level one heading.

The introduction ends with a clear statement of purpose.

Figure 11.10f. Revised Professional Formal Report in APA Style. This is Jessamyn's methodology section, which includes limitations.

The methodology section begins with a level one heading and explains how research was conducted.

You can use bullets to highlight important information, but don't exceed seven.

This subsection under Methodology begins with a level two heading. It recognizes the limits of available research. Not all reports require this section.

FEASIBILITY OF DRIVERLESS VEHICLES 5

Methodology

To weigh the actual costs vs. benefits of this option, the following areas must be considered:

- The annual operational cost of the current fleet in labor, routine maintenance, repairs, insurance, and fuel
- The initial cost of replacing the current vehicles with autonomous ones
- The estimated annual operational cost of the autonomous fleet in labor, routine maintenance, repairs, insurance, and fuel/energy
- The estimated gross revenues of the autonomous fleet vs. the known revenues of the current fleet, including estimated opportunity losses (lost revenue) for downed vehicles
- The safety risks and benefits to clients
- The potential regulatory changes or restrictions within the city and state

Limitations

This technology is new and while several companies are working to produce these vehicles, many are still in the development phase. There are few vendors for driverless vehicles at this time and the estimated cost to purchase, insure, and maintain them are estimated or theoretical.

Figure 11.10g. Revised Professional Formal Report in APA Style. This is Jessamyn's background section. It is more specific than the Introduction and can use background research.

Background

According to the U.S. Department of Transportation, 37,461 people were killed in traffic accidents in 2016 alone (National Highway Traffic and Safety Administration, 2019). Ninety-four percent of these crashes were the result of human error. Many companies hope to reduce these staggering statistics by shifting the work and responsibility of driving from the driver to the vehicle through an increase in automated features. In theory, fully autonomous vehicles will solve many of the problems that often arise from (or are at least made worse by) human error, such as accidents, traffic congestion, and wear and tear. There is even hope that automation will decrease fuel consumption, leading to reduced costs and lower greenhouse gas emissions (National Highway Traffic Safety Administration, 2019).

Most new vehicles already offer options designed to lessen some of the driver's burden. Various sensors, cameras, and other high-tech systems are in place to help the person detect objects in their blind spots, park the vehicle, and stay in their driving lane. Some are even able to steer and brake in an emergency. The National Highway Traffic Safety Administration (NHTSA) highlights six levels of automation as established by the Society of Automotive Engineers, with zero being no automation and five being fully automated (see Figure 1).

This signal phrase introduces the main government agency and the parenthetical citation provides the government agency as the author of this information.

This figure reference directs readers to the figure located at the end of the report.

Figure 11.10h. Revised Professional Formal Report in APA Style. Jessamyn's report includes definitions of terms.

Formal reports define essential terms. Notice how this single-sentence definition expands into a longer definition.

Define unfamiliar terms and acronyms by providing parenthetical definitions immediately afterwards.

This direct quote is followed by a parenthetical (in-text) citation. APA citations generally follow this format: (Author, Date, Page if any).

FEASIBILITY OF DRIVERLESS VEHICLES 7

Definition of Autonomous Vehicles

An autonomous vehicle (AV), also called driverless, is one that can operate without a driver present (level five). They accomplish this feat through a large array of sensors, cameras, global positioning systems (GPS), and a mix of software and hardware intended to replicate (or improve upon) human drivers. The most important features are the sensors. They serve as the eyes and ears of the vehicle, telling the body when and how to react. One of the most commonly used is light detection and ranging (LIDAR). This is a "remote-sensing [method] that can use light in the form of a pulsed laser to measure ranges" (Husain, 2017, p. 33). According to Waymo, a self-driving technology company, the sensors and software are "designed to detect and predict the behavior…of all road users," including cyclists and pedestrians. The vehicle is expected to know its exact proximity to any object at any time, the speed limit, where it is, and where it's going.

Figure 11.10i. Revised Professional Formal Report in APA Style. This is Jessamyn's findings section where she will put the majority of her research for the formal report. Formal reports use a blend of quantitative and qualitative research.

FEASIBILITY OF DRIVERLESS VEHICLES   8

Findings

Cost of Autonomous Vehicles

Today, the cost of driverless vehicles is significant, ranging from $225,000 to $1,000,000. The software and hardware packages are the costly portion, ranging from $70,000 to $150,000 (Lienert, 2017). According to Kevin Clark, the CEO of Delphi Technologies, "the cost of that autonomous driving stack by 2025 will come down to about $5,000 because of technology developments and (higher) volume" (as cited in Lienert, 2017). The cost of LIDAR is already declining. Velodyne, the primary producer of this technology, indicates that it will cut the $8,000 cost of its most popular sensor in half. Given that one vehicle likely has more than one sensor, the cost reduction is significant (Davies, 2018). The industry is aware that the high prices are a significant barrier to advancement (Davies, 2018).

Maintenance

Many of the vehicles are current models that have been modified, such as the Toyota Prius, Lexus RX450H, and Chrysler Pacifica. The sensors and parts to make it autonomous are developed by and purchased from various tech companies. With a modified vehicle, the parts and repairs for the vehicle are similar to what we see now, with the notable exception of

Notice the different heading levels on this page. APA style recommends this format for level one and level two headings.

This parenthetical citation shows how to cite an indirect source, which means that someone other than the author is cited.

Figure 11.10j. Revised Professional Formal Report in APA Style. Jessamyn's report addresses ethical concerns regarding her topic.

Notice how this sentence uses parallel structure to list actions: ride, idle, (do not) maintain, and park. The sentence that follows stands out because it is short and direct.

An ethical presentation of data will not ignore potential problems that might result from the proposed solution.

FEASIBILITY OF DRIVERLESS VEHICLES   9

the hardware and software specific to the driverless package. Other vehicles that are developed from start to finish primarily by the tech companies will be more unique, and access to parts will be more limited. The driverless vehicles are expected to save maintenance costs in some ways just by removing the driver. Humans ride their brakes, idle unnecessarily, do not maintain a consistent (or often legal) speed, and park too close to their neighbor. AVs avoid these issues.

Safety

As previously mentioned, AVs are expected to be safer than current vehicles. Humans are prone to distraction and overreaction. They don't react when they should or, in theory, as quickly as a computer. AVs are programmed to follow the rules of the road. They do not speed up when a light turns yellow, cut people off in traffic to make a turn they almost missed, forget to use their blinkers, text while driving, drive under the influence, or suffer from road rage. When all of the reasons accidents happen are considered, it is easy to understand why AVs could reduce accident statistics and save lives.

Concerns and Controversies

Many people, customers and drivers alike, fear the reliance on software. Many companies have had similar concerns that their technology

Figure 11.10k. Revised Professional Formal Report in APA Style. Jessamyn's report includes alternate points of view and credible research.

FEASIBILITY OF DRIVERLESS VEHICLES   10

may not yet have "learned" enough to be self-sufficient. They've chosen to keep a driver in the vehicle to take over in an emergency and the technology requires this (level three or four). This idea to keep the human in the driver's seat misses the main reason for wanting to remove them in the first place. Tesla, one of the companies that has chosen to form their own technology in lieu of LIDAR, uses a system they've named "Autopilot." In May 2016, a Tesla sedan ran into the broad side of a tractor trailer while the Autopilot was engaged. "Tesla said that Autopilot didn't register the white side of the trailer against the bright sky" (as cited in Simonite, 2016). Neither Autopilot nor the driver braked.

Other companies, like Waymo (Google's company) and NAVYA, have been operating their vehicles without a driver for some time now and have advanced their technology significantly. The cars have been in accidents, though few, and nearly all were the result of cars with drivers who didn't follow the rules. As these vehicles become more commonplace, which will be several years yet, these types of accidents should rarely occur.

Market Analysis

Pew Research Center recently surveyed over 4,000 Americans to determine their comfort and concerns with advances in technology, including driverless cars and technologies that would replace human

This sentence shows how you can insert your own voice into formal report without using the first-person perspective.

Credible research is important. Pew Research Center is well-respected. The sample size (4,000 Americans) is large enough to provide meaningful results.

Figure 11.10l. Revised Professional Formal Report in APA Style. Jessamyn cites her sources and includes a variety of sources in her report.

This parenthetical citation shows how to cite a source that has two authors.

Surveys can be an effective way to quantify the subjective experiences.

The parenthetical citation shows that this information comes from another source cited in the article by Smith & Anderson.

FEASIBILITY OF DRIVERLESS VEHICLES   11

workers (Smith & Anderson, 2017). They learned that "Americans are roughly twice as likely to express worry (72%) than enthusiasm (33%) about a future in which robots and computers are capable of doing many jobs that are currently done by humans" (Smith & Anderson, 2017). The majority believe that such advances will create a larger income inequality and that "the economy will not create many new, better-paying jobs for humans if this scenario becomes a reality." With specific regard to autonomous vehicles, 39% expect the roads will be safer while 30% think the vehicles will make them less safe. In fact, 56% indicated that they would not ride in a driverless car (as cited in Smith & Anderson, 2017).

Regulations

The regulations regarding automated driving systems (ADS) are not prohibitive. The NHTSA has established guidelines for states to help them safely integrate these vehicles onto their roadways. Most states require an application process for test vehicles, as well as licensing and registration that include more detailed information on vehicle ownership.

Regulations will likely adjust to state and federal levels as the technology progresses and as our government begins to realize that auto and tech companies looking at their bottom lines may not always be relied upon to move forward with caution or with the public's safety in mind. There

Figure 11.10m. Revised Professional Formal Report in APA Style. Jessamyn's report includes complex data and uses tables to make the information easier to understand.

are suspicions that some have moved forward with pilot programs even when they knew of flaws in their systems and that a lack of rules is to blame (Wolverton, 2018). The NHTSA wants this technology to move forward, as many do. Yet it is hard to regulate what is not yet understood, and, in this case, it is hard to fully understand it until tested under true conditions.

Cost Comparison

If Tomorrow's Taxi Company were to replace all vehicles, the numbers would look tentatively like the following:

	No. Cars	Driver/Operator		Fuel/Energy	Maintenance	Insurance	Revenue	Net
		No.	Annual Cost					
Current Fleet	40	65	2,710,000	237,104	50,536	151,609	6,551,020	$3,401,771
AV Fleet 20XX	38	10	360,000	94,605	40,858	163,435	7,062,020	$6,403,122
AV Fleet 20XX	38	8	380,000	94,605	34,049	149,815	7,062,020	$6,503,551

Table 1: Estimated Revenue in Dollars for an Autonomous Vehicle Fleet

This would be in addition to the purchase price of a complete fleet of driverless cars, which today would be very costly, approximately $9,400,000 for a fleet equaling the current size, plus the monitoring equipment.

Driver/Operator: Numbers include annual salary, benefits, and worker's compensation insurance for the drivers. Operators will replace the drivers, taking on the role of dispatching and monitoring the fleet remotely.

The formal report does not avoid complexity. The entire purpose of a formal report is to analyze a topic in depth.

Information is fictionalized for this table. All costs, unless otherwise cited, are also fictional. Your research, on the other hand, should be real.

Provide a title for all tables and visuals.

Figure 11.10n. Revised Professional Formal Report in APA Style. Jessamyn's report uses bullets and numbers to emphasize important information.

Lead into a list with a complete sentence followed by a colon.

Use numbers and bullets to emphasize important information and make it more visually appealing.

FEASIBILITY OF DRIVERLESS VEHICLES 13

Fuel/Energy: Tomorrow's Taxi Company's current vehicles are hybrids that average 50 MPG, driving approximately 2,339,000 miles per year, at an average of $2.90 per gallon. The AV fleet is electric and will get the equivalent of 93 MPG at a rate of $3.48 per 100 miles. The electric fleet should be able to drive approximately 2,522,150 miles per year.

Maintenance: Numbers for the current fleet are actual. Numbers for the AV fleet assume 3% of the IRS mileage reimbursement rate. This number reduces to 2.5% in 2021 with the assumption that the cost for parts, which are now quite expensive, will drop as mentioned. AVs are also expected to experience less wear and tear in most areas. This assumes the following:

1) Human error causes most accidents that result in repair.
2) Inconsistent driving practices, such as braking patterns, unnecessary idling, and not maintaining speed will be reduced in an AV.

What isn't clear, is the cost for parts and ease of replacement. While they are likely to spend less time in the service bay, they could be costly once there.

Insurance: This number is unknown for the AV fleet. Here it is estimated at the same rate of 12% of the IRS mileage reimbursement rate

Figure 11.10o. Revised Professional Formal Report in APA Style. The discussion section is where Jessamyn will analyze the sources she used in the formal report.

FEASIBILITY OF DRIVERLESS VEHICLES   15

sickness, and the general inattention humans have that software doesn't. The estimated annual amount of lost revenue from driver-related issues is approximately $30,600. Deloitte, one of the country's leading legal and financial consulting firms, has witnessed changing trends in the automotive industry. In an analysis of *The Future of Mobility Trends*, sees an expansion in what they call the "mobility ecosystem":

> Across the ecosystem, from auto retailers to insurance to finance, businesses are watching automotive industry trends and realigning to remain competitive and viable as the future of mobility unfolds. Companies that are preparing now—deciding where to play, transforming operations, implementing new technology, refocusing talent and marketing—will be well positioned to win. (Deloitte, n.d.)

This sentence shows how new information (the business name) is immediately defined within the sentence.

Direct quotes that exceed forty words should be formatted as a block quote without quotation marks.

FEASIBILITY OF DRIVERLESS VEHICLES   14

as it would be for a traditional vehicle. The numbers are expected to be lower as the technology becomes more established.

Revenue: These numbers are based on a rate of $2.80 per mile for both the current and AV fleet.

Net: This number is related to the direct cost of operating the vehicles and does not include numbers for any other operating costs for the company.

Discussion

The largest savings is in operational labor. The company would need to maintain a dispatcher and a small staff to monitor the position and condition of the vehicles. By removing the drivers, the company not only lowers the costs of annual salaries (an average of $34,000 per driver) but also benefits and worker's compensation insurance.

There is also a significant reduction in fuel. On average, drivers spend 30% of their time driving their routes looking for customers. The AVs

The discussion section is where you draw conclusions from your research and connect it to the issue you identified.

Figure 11.10p. Revised Professional Formal Report in APA Style. Jessamyn provides recommendations that include several options.

FEASIBILITY OF DRIVERLESS VEHICLES   17

would have the option to negotiate a purchase of the vehicles at a discounted rate.

Option Two – Test Vehicles

Under this scenario, Tomorrow's Taxi Company would lease-to-own two driverless vehicles from "Corporation A" at a rate of $20,000 per month for a period of six months. For the first six months, the corporation will handle the monitoring of the vehicles remotely. After the trial period, the

FEASIBILITY OF DRIVERLESS VEHICLES   16

> Recommendations may come before the conclusion in some reports.

Recommendations

Given the relatively small size of the company, and its financial limitations in comparison to industry leaders, Tomorrow's Taxi Company would need to tread carefully into the area of driverless vehicles. The autonomous projects that require a driver's presence have been excluded from the discussion for safety and cost reasons. The following options are

> A feasibility report ends with an analysis of options.

the most feasible at this time:

Option One – Partnership and Pilot

> Not all formal reports will provide multiple options. Sometimes, a single option emerges as the best choice.

"Corporation X" seems to be the U.S. automotive company most eager to release a fleet of AVs. They've developed partnerships with a few other companies in order to pilot their vehicles and are interested in using Tomorrow's Taxi Company as well. For the first year, they would provide the thirty-eight vehicles. Tomorrow's Taxi Company would be responsible for maintenance and energy costs, insurance, and monitoring staff. They would oversee any issues with the vehicles, make changes and upgrades, provide parts, and train our staff. They would receive .5% of fare profits for this period for the use of the vehicles. At the end of the contract period, it could be renegotiated for another year if the technology is still needing improvement (which would likely be the case) or Tomorrow's Taxi Company

Figure 11.10q. Revised Professional Formal Report in APA Style. This is Jessamyn's conclusion for her formal report. The conclusion typically includes the benefits of your recommendation.

Conclusion

The company will need to be prepared to make the transition to fully autonomous vehicles, likely within the next five years. The companies currently entering this field are large players in the automotive or technology industries, with the knowledge and financial resources to venture into this unchartered territory. Driverless vehicles are still a work in progress, though the progress is impressive. They are a new and exciting technology but also a frightening one for many. For that reason, they are under a microscope with various authorities and the community in general, and a company like Tomorrow's Taxi Company may be highly vulnerable should one of the vehicles have a safety issue. However, the taxi industry will be the first to integrate AVs regardless of such concerns, so it is important that the company move forward proactively but cautiously.

The concerns with the technology over safety, regulatory changes, and cost are there; however, at the current pace, these will soon be of little issue. The AVs, though not without their accidents, are still safer than Tomorrow's Taxi Company's current vehicles. As the numbers are tracked, studies are done, and more people experience the technology, a company with a driverless fleet will find itself well-positioned to succeed.

Conclusion sections that follow the recommendation section will discuss the benefits of your recommendation.

End with a forward-thinking call to action.

Figure 11.10r. Revised End Matter in APA Style. Jessamyn's report ends with a list of references for her research.

This entry shows the standard format for an electronic source:

Last Name, F. (Year, Month Day). *Title of page.* Site name. URL

In this entry, the author is the company. When no publication date (n.d.) is provided, list the date you accessed the site.

This entry shows the standard format for a book:

Last Name, F. (Year). *Title: Subtitle if any.* Publisher.

For government documents, use the more specific agency name as the author. List the parent agency as the publisher before the URL.

FEASIBILITY OF DRIVERLESS VEHICLES   20

Smith, A., & Anderson, M. (2017, October 4). *Automation in everyday life.* Pew Research Center. http://www.pewinternet.org/2017/10/04/automation-in-everyday-life/

FEASIBILITY OF DRIVERLESS VEHICLES   19

References

Davies, C. (2018, January 2). *A key part of many autonomous cars just got a huge price cut.* Slash gear. https://www.slashgear.com/velodyne-lidar-puck-autonomous-car-more-affordable-02513340/

Deloitte. (n.d.). *Future of mobility trends.* Retrieved April 4, 2021, from https://www2.deloitte.com/us/en/pages/consulting/solutions/future-of-mobility-trends-industry-ecosystem.html

Husain, A. (2017). *The sentient machine: The coming age of artificial intelligence.* Scribner.

Lienert, P. (2017, December 5). *Cost of driverless vehicles to drop dramatically: Delphi CEO.* Insurance Journal. https://www.insurancejournal.com/news/national/2017/12/05/473134.htm

National Highway Traffic Safety Administration. (2019). *Automated vehicles for safety,* U.S. Department of Transportation. https://www.nhtsa.gov/technology-innovation/automated-vehicles-safety

Simonite, T. (2016, June 30). *Fatal Tesla's Autopilot crash is a reminder autonomous cars will sometimes screw up.* MIT technology review. https://www.technologyreview.com/s/601822/fatal-tesla-autopilot-crash-is-a-reminder-autonomous-cars-will-sometimes-screw-up/

Figure 11.10s. Revised End Matter in APA Style. Figures and other additional material are often included in the report's end matter.

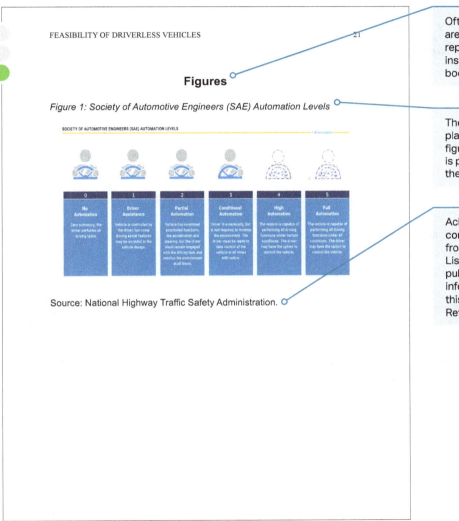

FEASIBILITY OF DRIVERLESS VEHICLES   21

Figures

Figure 1: Society of Automotive Engineers (SAE) Automation Levels

Source: National Highway Traffic Safety Administration.

Often figures are placed in the report's end matter instead of in the body of the report.

The figure label is placed above the figure. The source is provided beneath the figure.

Acknowledge content and visuals from other sources. List the complete publication information for this source on the References page.

Checklist for Revision

Features

- [ ] Does the working title still reflect the report's revised content?
- [ ] Are the headings specific to each section's content and consistently styled using three levels of headings?
- [ ] Is each visual communicating a key message in the most effective way—bulleted lists, tables, graphics, etc.—and with a consistent visual style?

Material

- [ ] Does the introduction set up the context with enough background, discuss any methodology or limitations, and present a guiding purpose statement?
- [ ] Does each section in the body stay focused on one relevant topic?
- [ ] Do the recommendations describe the option(s) in enough concise detail so the outcome can be fulfilled in a timely, practical, and efficient way?

Requirements

- [ ] Are all sections in the report listed correctly in the table of contents?
- [ ] Is every source accounted for in your document, in your visuals, and in your bibliography?
- [ ] Is the letter of transmittal properly addressed?

Conclusion

Formal reports draw on all the skills explored in this textbook and allow you to bring all of your expertise together to create a successful finished document. Your use of multimodal strategies, visual elements, effective definitions, concise descriptions, accessible instructions, convincing proposals, and analytical formal reports can get you closer to providing a solution.

Jessamyn's report ultimately led her boss to refrain from the purchase of AVs for now, but it earned her a promotion. She's now known within the company for her capacity to conduct research, analyze information, and clearly communicate complex ideas. Your future is equally bright. Even when cars can drive themselves, technical communication will always need a human touch.

Moving from problem to solution is not a process you'll complete just once. If you're good at it, you'll continue to find communication challenges throughout your career. These challenges are opportunities to deliver helpful messages that will benefit others and bring you satisfaction in your work.

Chapter 12

Make Technical Communication Work for You

Abstract: Not everyone becomes a technical writer, but most of you will use technical communication in your chosen fields. At its core, technical communication is a specific message with a clear purpose for a certain audience. Strong communication skills are essential to success in any field you choose. Everything you write is a reflection of you and the company or organization you represent. And it's not just about you—collaboration is also part of technical writing. Learning how to bridge the gap between student and professional work can be useful as you advance in your education and career. This chapter explores job postings that require technical communication skills. This chapter also offers practical advice from a diverse group of professionals who have found ways to make technical communication work for them.

Looking Ahead

1. Bridging the Gap
2. Types of Communicators
3. Meet the Professionals
4. Tips for Success

Key Terms

» freelancer/contractor
» network
» technical communicator
» technical writer

Bridging the Gap

How your undergraduate classes connect to each other and your profession will inform your educational path. For example, once you see how this technical communication course helps you write an individualized educational plan for one of your future students or how a sociology course helps you become a better manager, then you have begun synthesizing your education with your professional life. Some elements of your undergraduate work may never apply professionally, and other elements will surprise you with their relevance. Regardless of your path, strong communication skills are fundamental to all higher-wage postsecondary careers.

From the Classroom

Educational requirements and standards in the classroom are set by your professors. They tell you what to write, how much, and when it's due. They model what to do and ask you to practice before you perform the task. You're often asked to explain the choices you made and how you got to the answer. If you stumble or fall, your professors are there to show you how to learn from the experience and rise to your feet.

This type of guided instruction prepares you for the workforce because it gives you something to lean into and trust. It also teaches you the flexibility and curiosity to adapt to new expectations and solve problems. Institutions of higher learning invest money and resources so that you can be successful. Worst-case scenario—you don't pass a class. Even with these guardrails in place, it's important to establish good habits to carry with you beyond the classroom.

To the Professional World

Professional demands vary widely. The principles taught in this book aim to encourage discipline and resourcefulness as you continue to learn skills for technical communication. It's important to tap into and trust your resources because you'll need many of these skills as you train for new positions and accumulate professional experience.

You may have noticed that this book focuses more on longevity skills like writing with clarity and understanding design instead of using specific software. That's because software advances so quickly and companies often choose different programs. Instead of giving you step-by-step tutorials on how to use specific programs to create documents, we provide you a foundation to work from and build upon.

Some lessons from your technical communication course are universally applied, others will need to be modified to fit the specifics of your workplace, and a few will become surprisingly relevant years from now. Your education enhances your workplace training and gives you the flexibility that comes with having the right tools for the right situations.

Types of Communicators

The following sections describe the different kinds of work that require technical communication skills. Not everyone who studies technical communication will become a technical writer, but most people will benefit from knowing how to communicate clear, accurate, and precise information to a specific audience.

Technical Communicator

Technical communicator is a specific job that clearly uses the skills from this textbook, along with those you might learn if you major in technical writing. But the places in which you can expect to use technical communication daily might surprise you. Technical communicators are people working in fields where they communicate specialized knowledge. Most did not major in writing or communications, but all use those skills in their fields. The Bureau of Labor Statistics (BLS) reports that the highest-paying postsecondary certificate careers require professional-level communication skills, including the following positions:

» Electrical and electronics repairers, powerhouse, substation, and relay
» First-line supervisors of firefighting and prevention workers
» Insurance appraisers, auto damage
» Aircraft mechanics and service technicians

This list shows some of the higher-paying careers you can enter by earning the appropriate certificate. The job posting in figure 12.1 requires technical communication skills, even though the job involves repair of electronic equipment.

The following list is, according to BLS, associate's degree-level careers with median wages above $70,000 per year, all of which include communication skills as a requirement:

» Dental hygienists
» Radiation therapists
» Diagnostic medical sonographers

The following six-figure careers typically require a bachelor's (or higher) and, you guessed it, require excellent communication skills and, in some cases, the ability to write and publish research:

» Computer and information systems managers
» Architectural and engineering managers
» Petroleum engineers
» Nurse practitioners
» Physician assistants
» Nurse anesthetists
» Psychiatrists

Figure 12.1. Technical Communication Skills Required. Even though "technical communicator" is nowhere near the job title, notice how communication and technical skills are part of the job's requirements.

Preferred Attributes and Qualifications
- <mark>Degree and or certification from college/technical/ vocational school.</mark>
- Years of related experience: Over one year and up to three

Job Title: Electronic Repair Technician
Location: Portland, OR

Position Summary
Tests, modifies, upgrades, performs basic repairs, calibrates, aligns, and burns-in assigned client computer hardware to the component level following established procedures and guidelines.

Position Responsibilities and Essential Functions
- Tests, repairs, burns-in, and cleans assigned equipment in accordance with policy.
- Interacts on a professional level with associates, participating in job circles, cross-training, and information sharing.
- Monitors assigned equipment safety stock levels and maintains enough equipment to meet safety stock requirements.
- Maintains work area and assigned test equipment and tools.
- Follows department static control procedures.
- Maintains required paperwork, records, and documentation.
- Performs other related duties as assigned.

Qualifications and Requirements
- Good mechanical skills including soldering skills.
- Ability to use troubleshooting strategies.
- Ability to lift up to fifty pounds.
- <mark>Ability to work effectively in a team environment.</mark>
- <mark>Communicates effectively verbally and in writing.</mark>
- Ability to adapt to shifting priorities and be able to handle multiple tasks simultaneously.
- Must be able to demonstrate an understanding of basic electronic theory.

Be prepared to talk about your educational experiences in a job interview.

Repair and technician work often requires updates, recommendations, estimates, and organizing with coworkers and clients. This is why you typically see communication listed under requirements instead of preferences.

Technical Writer

A **technical writer** is someone who creates instruction manuals, informational reports, and other documents communicating complicated technical information to specific audiences. These are people who have chosen to use their writing skills as technical writers. Most majored in related fields like English, media, or communications. However, enterprising engineering students have learned to minor in technical writing as a strategy to stand out in competitive fields.

You can also choose to major in technical writing itself, with a minor in science, engineering, or other technical fields. Whether or not a technical writer majors (or minors) in a technical field, they mostly write for engineering and computer industries and need to be familiar with these industries. This is where collaboration plays a part. It takes a team of professionals to create a balanced, useful document.

In this next job posting, notice all the highlighted areas that align with principles from this textbook (figure 12.2). What other connections do you see with the course material?

Freelancer/Contractor

One of the realities of living in the twenty-first century is the rise of the "gig" economy. Gigs come in all shapes and sizes, from ridesharing jobs to listing how much work you'd do for $5. But there is one commonality among these jobs: they are all freelance/contract work. **Freelancers**, sometimes called **contractors**, are self-employed individuals who perform duties defined in a contract with an employer or client.

Temporary, contract, and freelance labor is attractive to organizations because the people filling these posts are less expensive than full-time, permanent workers. These temporary workers aren't considered employees in the traditional sense, and this means they often work from a home office or other rented space. While this type of work seldom comes with benefits like health insurance, some people like this style of employment for the flexibility and the chance to work in a range of fields.

Figure 12.2. Most Jobs Require Communication Skills. Scan this job description, paying special attention to the highlighted portions. Notice how the requirements of this position align with the content covered in this textbook.

- Excellent communication and collaboration skills (both written and verbal).

Preferred Qualifications:
- Technical certification(s) in networking area

About the Position:
The Information Experience (iX) team develops and delivers the content our customers need to install, manage, and configure Juniper products. The iX team strives to deliver the right product information in the appropriate medium to meet customer requirements and exceed their expectations. The iX team is looking for a motivated candidate to create product documentation, including comprehensive concepts, examples, tasks, and reference information for Juniper's routing, switching, and security products.

Responsibilities:
- Work with Engineering, System Test, Program Management, Customer Support, Product Line Marketing, and other key stakeholders to identify requirements and develop best-in-class documentation.
- Incorporate feedback from customers and internal sources to improve existing and future documentation.
- Work with editors and other writers to identify and implement standards and process improvements to enhance product documentation and usability.
- Participate in special projects or initiatives outside the scope of regular tasks.
- Follow established style and process guidelines to provide consistency and completeness.

Minimum Qualifications:
- Bachelor's degree, Entry-level position with one to two years of technical writing experience and/or previous internship.
- Knowledge of networking technology areas, such as switching, routing, and security.
- Ability to transform complex technical concepts and specifications into easily understood, clear tasks and concepts.

This is the Problem-Solution Framework in action.

Revision is not just something you do in school.

This is what technical communication is all about.

Contract or freelance work can be exciting, but this type of work comes with many feast or famine cycles. A successful freelancer knows how to save their money for the inevitable slow period between gigs. Inconsistent employment is the norm for freelancers. Another major factor to consider is that most freelance work requires self-motivation and discipline to find contracts and then to do the work. The ability to realistically assess and communicate the scope of a project is vital to successful freelancing.

Freelancing often makes it possible for people to work who might not otherwise be able to work because of personal circumstances or to supplement their incomes from traditional employment. According to "Freelancing in America: 2019," a comprehensive measure of independent workers, 57 million Americans freelanced in 2019, which represents about 35 percent of the U.S. workforce. You may hear varying reports about whether the gig economy is growing or shrinking because it's hard to accurately track freelancers and contractors. Additionally, many positions come with a variety of job titles that confuse data collection. For example, your "professor" (a specific job title) might actually be an "adjunct" who is officially titled as "part-time instructor," which is actually a temporary contract position that renews every quarter or semester. On paper, it may look like the college hires a bunch of permanent part-time faculty, but the reality is completely different.

According to BLS, the following types of freelance gigs are more common:

» Arts and design (ranging from graphic design to crafts and fine arts)
» Computer and information technology (web and software designers, programmers)
» Construction and home repair
» Media and communications (tech writers, interpreters, translators, and photographers)
» Transportation (ride-sharing drivers, delivery drivers)

Even if you decide that freelancing full-time is not for you, short-term contracts and freelance assignments can bring in extra money. The following example is for a graphic design freelance position at a regional radio station (figure 12.3). Sometimes people are surprised at how much technical communication is expected as part of "non-writing" careers.

Figure 12.3. Job Description for Technical Communicator. Notice how this graphic design job positing requires many of the skills you've learned in this textbook, and not just in visual design principles.

- Knowledge of digital and print media standards and practices.
- Ability to work within a budget.
- A degree in Graphic Design or related field is preferred.

Job Description

All Classical Portland seeks a creative, enterprising graphic designer. The ideal candidate will be part of a collaboration where your creativity, personality, and integrity will help shape the image and branding of a dynamic radio station, helping to carry out All Classical Portland's mission to build cultural community and provide access to the arts for all. This position will conceptualize, design, and lay out a wide variety of materials for internal and external use, including promotional materials and multimedia campaigns. You will act as a liaison between internal project managers and external vendors, obtaining quotes and preparing projects for outside printing companies, while keeping the internal project manager apprised of the status throughout the production process. This position is a great opportunity for fun, creative artistry. Candidates should have superior multitasking and organizational skills and be able to meet deadlines.

Essential Responsibilities
- Conceptualize, design, and lay out a wide variety of materials for internal and external use.
- Collaborate in planning and designing an annual and monthly marketing strategy.
- Work with the Community Engagement and Management teams to create a streamlined brand identity.

Required Qualifications
- At least three to five years of design experience and a strong understanding of design fundamentals, branding, typography, and visual hierarchy in layout.
- Proficient in graphic design and layout; creative in finding solutions to design needs.
- Demonstrated ability to work both independently and collaboratively under deadline pressure.
- Proven intellectual curiosity, creativity, and rigor with strong writing skills.

This job requires you to be able to adapt your communication style to different types of audiences.

Get ready to write more status reports.

Effective writing skills are essential in all careers.

Meet the Professionals

The professional profiles collected in this chapter represent those who have bridged the gap between their education and their career. We sent out a survey to a variety of professionals who use different types of communication in their field. Note the common skill sets and the areas where flexibility comes into play.

Technical Communicator Profile

These are profiles of people working in fields where technical communication is a significant component of the job duties. Most did not major in writing or communications, but all use those skills in their fields.

Q & A with a Head of Products and Services

Q. *How do you explain your job to someone outside your field?*

A. [I] create and tell the story of our brand through useful products.

Q. *What technical documents do you feel are relevant to educating a future professional in your field?*

A. User story definition, acceptance criteria, Product Requirements Doc (PRD), Marketing Req Doc (MRD), Pitch Decks, term sheets, and machine-readable resumes.

Q. *What do you wish you'd had access to as a student training for your profession?*

A. A clue, some mentorship.

Q. *What's the biggest difference between academic writing and professional communication in your field?*

A. Writing clear, thoughtful narratives are more important in tech writing than was previously taught. The documents of yesteryear (memos, spec docs) have disappeared.

Q & A with a Designer

Q. *How do you explain your job to someone outside your field?*

A. [I] design underground storm and sanitary sewers.

Q. *What technical documents do you feel are relevant to educating a future professional in your field?*

A. The Engineers Joint Contract Documents Committee (EJCDC) Contract Documents. Understanding the front-end documents to my project and the contract itself is extremely important to understand construction and the responsibility in construction.

Q. *What do you wish you'd had access to as a student training for your profession?*

A. Construction Document Technologies (CDT). I have recently completed this and it would have been beneficial in college.

Q. *What's the biggest difference between academic writing and professional communication in your field?*

A. In college, it was stressed to keep it to the point and technical. I have found that as a professional, you have to adjust your writing to your audience. I have to explain items differently to a homeowner than I do a colleague. They only really teach the technical, but I have found that explaining in everyday terms is just as important.

Technical Writer Profile

These profiles are people who have chosen to use their writing skills as technical writers. Most majored in related fields like English, media, or communications.

Q & A with a **Principal Technical Writer**

Q. *How do you explain your job to someone outside your field?*

A. [I] write documentation (instruction manuals) for how to use my company's software with multiple audiences in mind.

Q. *What technical documents do you feel are relevant to educating a future professional in your field?*

A. There's so many! *The Product Is Docs* [a book by the Splunk documentation team], the *Write the Docs* Slack [an online network of professionals], and *Every Page Is Page One* [a book by Mark Baker] for starters.

Q. *What do you wish you'd had access to as a student training for your profession?*

A. Honestly, I wish I'd had a mentor. I probably would have jumped from support into writing sooner.

Q. *What's the biggest difference between academic writing and professional communication in your field?*

A. Having an actual editor, plus the pressure of knowing that if I mess something up, I could partially be responsible for causing a customer issue.

Q & A with a Technical Writing Instructor and Technical Writer

Q. *How do you explain your job to someone outside your field?*

A. Technical writing is clearly explaining literal information in a way that makes it perfectly understandable and usable to its intended audience.

Q. *What technical documents do you feel are relevant to educating a future professional in your field?*

A. Procedures, emails, functional descriptions, technical illustrations, specifications (both "what it is" and "how it should work"), troubleshooting trees, quick reference cards, formal reports, and articles.

Q. *What do you wish you'd had access to as a student training for your profession?*

A. An instructor who had actually done the job! Great examples of good work and explanations of how bad work specifically didn't measure up. Knowledge of word processor features like outlining, automatic table of contents, document element styles (e.g., XML tags) vs. formatting, and a really good peer reviewer.

Q. *What's the biggest difference between academic writing and professional communication in your field?*

A. There should be none, but academic writing tends to be ego-driven in tone: more words, more syllables, longer and more cumbersome sentences, incomprehensible vocabulary, and gibberish titles seem to be the rule. Also, academic work has a much higher emphasis on citing all sources both in text and in a bibliography; technical work may give an acknowledgment somewhere or do nothing at all.

Unexpected Technical Communicators

These are profiles from others who use technical communication in unexpected ways.

Q & A with a Real Estate Broker

Q. *How do you explain your job to someone outside your field?*

A. I help people buy and sell homes.

Q. *What technical documents do you feel are relevant to educating a future professional in your field?*

A. The online courses for obtaining a license and for continuing education are horribly written. I think they were written by an attorney—very confusing language.

Q. *What do you wish you'd had access to as a student training for your profession?*

A. Maybe more real-world accounts of experiences/ scenarios as part of the [real estate] curriculum, written by actual agents (or writers), not attorneys. The material is so dry and hard to get through.

Q. *What's the biggest difference between academic writing and professional communication in your field?*

A. [Real estate] is a people-based one with plenty of face-to-face time and phone interaction, as opposed to academic documents.

Q & A with a **Finance Manager**

Q. *How do you explain your job to someone outside your field?*

A. I take care of financing vehicles for a dealership.

Q. *What technical documents do you feel are relevant to educating a future professional in your field?*

A. Excel [spreadsheets].

Q. *What's the biggest difference between academic writing and professional communication in your field?*

A. My professional communication is much shorter in comparison to academic writing.

Tips for Success

Education is a lifelong pursuit. It gives you the awareness and flexibility to meet changing job markets, technology, and to adapt accordingly. It also exercises your curiosity, which is a fundamental survival skill that keeps you marketable. Your training shouldn't stop at graduation.

Stay Current

Stay current in your field with continued education. Many careers require you to continue earning credits to maintain licensure. Even if your field doesn't have license requirements, education is a solid investment. You can attend industry conferences. Many of them offer credits for their proceedings. Look into area colleges and see what relevant classes or certifications are offered.

Become a member of industry-related organizations. Most industries have their own organizations. For instance, computer scientists and engineers can join IEEE (https://www.ieee.org). There are also associations for under-represented individuals as well. Women in Tech Comm, a special interest group from the Society for Technical Communication, is just one example. You will find the support and resources needed for a long and productive career.

Remember to **network** and stay connected to other professionals. There is truth to the adage "it's who you know" when it comes to developing new opportunities. Continued education, trade shows, conferences, and professional organizations provide networking opportunities. Just like in college, your best resource may be sitting next to you.

Pay It Forward

In the same way that the people profiled in this chapter took time out of their day to share their experience, you should do the same. There will be times when you have more time than money, so help out future professionals in your field by volunteering to share your knowledge. We learn from those who came before us. Go back to your college or university and speak to a new crop of young professionals—share your experiences. Become a mentor to someone. Sometimes, having one person believe in you makes all the difference.

Case Study

You: A Case Study

This textbook presents case studies for you to see concepts in action. Now, it's your opportunity to reflect on everything you learned and how you might apply these ideas to your work as a student and professional.

This chapter shows you the bridge between student work and the work you'll do beyond the classroom. A key component of education is learning how to synthesize your classes while in school and then learning how to integrate your education with your professional life. Reflection allows you the opportunity to put all of the pieces together so you can continue adding to the puzzle over the course of your educational pursuits and professional career.

Discussion

» Write a paragraph or two (at least 250 words) on what you learned this term. What stands out to you? What will be most helpful as you continue your education? Do you see connections between the content in your technical communication course and your other classes (current or future)? How can you use what you learned from this book in other classes?

» Write a paragraph or two (at least 250 words) about how to incorporate content from this class into your professional life. What skills are sought after in your field? How does communication play a role in your future profession? Go back and review the chapters, then connect skill sets to your profession.

» Write a paragraph or two (at least 250 words) about how you plan to stay current in your field. Will you continue your education? Are there certificates or licenses that you can pursue? What are your networking opportunities with others in your field? Look at related affiliations, organizations, clubs, etc.

Checklist for Your Future

Synthesize

- ☐ Reflect on how your educational training supports your professional goals.
- ☐ Articulate previous work, internship, and/or relevant volunteer experience.
- ☐ Create a goal statement that describes current and future goals.

Prepare

- ☐ Develop a list of key words for your desired field or profession.
- ☐ Look for companies you might want to work for and read their mission statements.
- ☐ Arrange an informational interview. Prepare at least three to four questions to show your interest in the organization.

Keep Learning

- ☐ Research certificates and licenses in your field so you can stay relevant.
- ☐ Join affiliations that allow you to meet new colleagues.
- ☐ Develop a plan to stay involved with new and emerging professionals.

Conclusion

Workplaces need employees who can think critically about information and design in this multimodal world. You can use the skills in this book to apply for jobs, communicate in the workplace, create and design ethical documents, and provide resolutions to problems that need them.

This book is an introduction to a precise and specific way of communicating that will serve you well throughout your career. Think of it as a building block, and look for other opportunities that will help you build invaluable tools such as communication, design, and innovation. Be open to learning and new encounters. Getting a job is not the end of your journey—it's the beginning. Remember your future employer is looking for a problem solver.

Acknowledgments

Practical Models for Technical Communication is the result of countless hours of faculty and student input. Faculty on the development team visited classes, surveyed students, and analyzed the content of all the leading technical communication textbooks.

This book has undergone three labor-intensive development editions since 2018 that involved additional research, writing, and design work. This collaborative process has been as instructive as the book itself, and Chemeketa Press is grateful for the expertise and dedication of who those have helped shape the book, from initial outline to the complete first edition.

We offer thanks to the following faculty for their contributions: Greg Berry, Shobana Breeden, Chris Cottrell, Liaken Hadley, Adam Karnes, Brian Mosher, Magdalen Powers, and Catherine Shride. We also appreciate those who field-tested the development editions of this book and/or provided feedback: Lani Davison, Matthew Hodgson, Layli Liss, Jean Mittelstaedt, Ruth Perkins, Suzanne Spencer, Cindy Ulshafer, Jan VanStavern, and Jaime Zinck. Thank you to those who offered feedback or reviewed the book's outline in its early stages of development: Aaron Bannister, Ryan Davis, Jeffrey McAlpine, and Peter Starr. Additionally, we want to recognize the support of the English department with special thanks to the WR227 work group, where some of the first conversations about the direction of this book took place.

Our students are the heart of everything we do, and we want to acknowledge a few by name who have made significant contributions to this project: Brandi Harbison, Nadia Isom, Shaun Jaquez, Cassandra Johns, Leo Martinez, Tiana Miller, Matt Sanchez, Brice Spreadbury, and Taylor Wynia.

This book would not have been possible without a grant from John and Bobbie Clyde. Their generous gift supported the involvement of faculty in this project.

The author thanks Don Brase and Kim Colantino for entrusting her with this project. She would also like to thank Leslie Kimiko Ward, a partner in all things, for her patience and creative support.

The author would also like to thank the editor, Stephanie Lenox, for her continual guidance and commitment to this book.

Image Credits

Chapter 4, figure 5. Used with permission from Tom Johnson's "I'd Rather Be Writing" podcast at (https://idratherbewriting.com/).

Chapter 8, figure 6. Student work is copyrighted and may not be used without permission from the publisher.

Chapter 9, figures 1, 10, and 13. Used with permission from Sauder Woodworking Co.

Chapter 9, figures 2 and 11. *Sexy Technical Communication* is licensed under a Creative Commons Attribution 4.0 International License at (http://distanceed.hss.kennesaw.edu/technicalcommunication/).

p. 261, Checklist for Document Accessibility; Courtesy of Heather Mariger and the Center for Academic Innovation, Chemeketa Community College, Salem, Oregon. Originally adapted from accessibility guidelines developed by Rondi Schei and Heather Mariger for Oregon Community College Distance Learning Association.

Appendix A

MLA Report

Figure A.1. This is Jessamyn's revised letter of transmittal.

November 15, 2020

Morgan Milford, CEO
Tomorrow's Taxi Company
1111 NE Rogers Dr.
Averagton, AZ 62000

Dear Mr. Milford:

Enclosed is my report, "Feasibility of the Implementation of Driverless Vehicles," for your review. Thank you for giving me this opportunity and allowing me to further my knowledge and the knowledge of our organization on this very important issue.

As you will see, some of this report's information is theoretical because driverless vehicles are not yet commonplace and are operated primarily by large automotive technology companies. Taxi companies are just beginning to join these ranks, and I'm confident that the material enclosed will help Tomorrow's Taxi Company decide when to make a similar leap.

I would like to thank our marketing, service, payroll, and human resources departments for contributing to this report. Their knowledge and data are essential to our decision. If you need any additional information, please contact me. I can be reached at *sanchez.j.@ttc.com* or 555-222-1000, ext. 200. Thank you for your time and consideration in reviewing this proposal.

Sincerely,

Jessamyn Sanchez
Jessamyn Sanchez
Director of Business Development

Get down to business with the first sentence of your letter of transmittal.

Show that you are a good colleague by naming the individuals and departments that contributed to your report.

A letter of transmittal should not be more than one page or three paragraphs in length.

Figure A.2. This is Jessamyn's title page for her formal report using MLA style.

The report's title should be in bold and centered. The first, last, and all other important words should capitalized.

Indicate who the report is being prepared for here. Include name, title, company, and location.

Indicate who prepared the report here. Include name, title, and company.

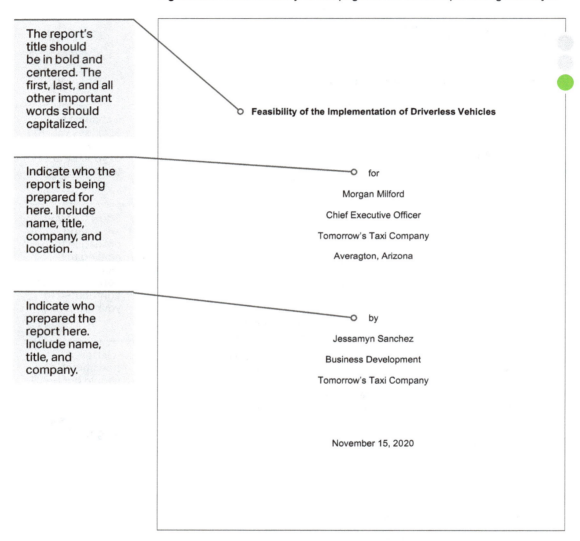

Feasibility of the Implementation of Driverless Vehicles

for

Morgan Milford

Chief Executive Officer

Tomorrow's Taxi Company

Averagton, Arizona

by

Jessamyn Sanchez

Business Development

Tomorrow's Taxi Company

November 15, 2020

Figure A.3. This is Jessamyn's table of contents for her formal report using MLA style.

The table of contents lists the headings and subheadings as they appear in the report.

Table of Contents

This report includes three levels of headings.

This report is written in MLA style, so it includes a Works Cited page.

Figure A.4. This is Jessamyn's executive summary, also called an abstract.

In student reports, the student's last name and page number should appear in the running header. In professional reports, the title of the report and the page number should appear in the running header.

The executive summary is sometimes called an abstract.

This executive summary is exactly 150 words. Most abstracts are about this length and presented as a single paragraph.

Feasibility of Driverless Vehicles 1

Executive Summary

Due to the age and high cost of maintaining its current fleet, Tomorrow's Taxi Company looks to determine the feasibility of replacing their current vehicles with autonomous vehicles (AVs). The technology is new and still developing, but recent advances mean the technology is becoming safer for customers and more cost effective to operate. However, AVs are not yet readily available and affordable. The primary players in this field are large automotive and technology companies with large amounts of capital and industry support, though they are beginning to offer their vehicles to other companies, such as taxi and delivery services. Clearly, the market is moving quickly in this direction. Costs, which are quite high at this time, are expected to become more competitive within the next few years. It is recommended that the company begin cautiously transitioning into the driverless market and preparing its business model and employees for this change.

Figure A.5. This is Jessamyn's introduction for her formal report. You generally do not find cited material in the introduction. It is your opportunity as the writer to introduce relevant information, in your own words.

Feasibility of Driverless Vehicles 2

Introduction

> The introduction provides relevant information for the report.

Over the past two years, Tomorrow's Taxi Company has experienced an increased financial burden arising from its aging fleet of vehicles. The concerns include an increase in expense for parts and labor, as well as lost revenue due to unavailable cars. Employee turnover has added to this cost. Drivers expect a consistent number of work hours and a certain level of tips, and many have left to work for other companies, leading Tomorrow's Taxi Company to more frequent hiring and training expenditures.

Many companies are experiencing this ongoing issue. Some businesses are considering the addition of driverless vehicles, which are promoted as more cost effective and less prone to issues that result from poor driving practices. Cities across the world have implemented pilot programs of both semi and completely autonomous fleets with the goal of cutting down on labor expenses, increasing safety (lower liability costs), and promoting a more technologically advanced (and often more environmentally sound) service to its consumers. This report investigates the economic feasibility of replacing Tomorrow's Taxi Company's current vehicles with an autonomous fleet.

> The introduction ends with a clear statement of purpose.

Figure A.6. This is Jessamyn's methodology section, which includes limitations.

The Methodology section begins with a level 1 heading and explains how research was conducted.

You can use bullets to highlight important information, but don't exceed seven.

This subsection under Methodology begins with a level 2 heading. It recognizes the limits of available research. Not all reports require this section.

Feasibility of Driverless Vehicles 3

Methodology

To weigh the actual costs vs. benefits of this option, the following areas must be considered:

- The annual operational cost of the current fleet in labor, routine maintenance, repairs, insurance, and fuel
- The initial cost of replacing the current vehicles with autonomous ones
- The estimated annual operational cost of the autonomous fleet in labor, routine maintenance, repairs, insurance, and fuel/energy
- The estimated gross revenues of the autonomous fleet vs. the known revenues of the current fleet, including estimated opportunity losses (lost revenue) for downed vehicles
- The safety risks and benefits to clients
- The potential regulatory changes or restrictions within the city and state

Limitations

This technology is new and while several companies are working to produce these vehicles, many are still in the development phase. There are few vendors for driverless vehicles at this time and the estimated cost to purchase, insure, and maintain them are estimated or theoretical.

Figure A.7. This is Jessamyn's background section using MLA-style citations.

Feasibility of Driverless Vehicles 4

Background

According to the U.S. Department of Transportation, 37,461 people

were killed in traffic accidents in 2016 alone (National Highway Traffic).

Ninety-four percent of these crashes were the result of human error. Many

companies hope to reduce these staggering statistics by shifting the work

and responsibility of driving from the driver to the vehicle through an

increase in automated features. In theory, fully autonomous vehicles will

solve many of the problems that often arise from (or are at least made

worse by) human error, such as accidents, traffic congestion, and wear and

tear. There is even hope that automation will decrease fuel consumption,/

leading to reduced costs and lower greenhouse gas emissions (National

Highway Traffic).

Most new vehicles already offer options designed to lessen some of

the driver's burden. Various sensors, cameras, and other high-tech

systems are in place to help the person detect objects in their blind spots,

park the vehicle, and stay in their driving lane. Some are even able to steer

and brake in an emergency. The National Highway Traffic Safety

Administration (NHTSA) highlights six levels of automation as established

by the Society of Automotive Engineers, with zero being no automation and

five being fully automated (see Appendix A).

This signal phrase introduces the main government agency and ends with a parenthetical citation that names the government agency (in a shortened form) as the author of this information.

The reference at the end of this sentence directs readers to the figure located in the Appendix at the end of the report.

Figure A.8. Jessamyn's report includes definitions of terms and MLA-style citations.

Formal reports define essential terms. Notice how this sentence definition becomes an expanded definition.

Define unfamiliar terms and acronyms by providing a parenthetical definition immediately afterwards.

This direct quote is followed by a parenthetical (in-text) citation.

Feasibility of Driverless Vehicles 5

Definition of Autonomous Vehicles

An *autonomous vehicle* (AV), also called *driverless*, is one that can operate without a driver present (level five). They accomplish this feat through a large array of sensors, cameras, global positioning systems (GPS), and a mix of software and hardware intended to replicate (or improve upon) human drivers. The most interesting and important features are the sensors. They serve as the eyes and ears of the vehicle, telling the body when and how to react. One of the most commonly used is light detection and ranging (LIDAR). This is a "remote-sensing [method] that can use light in the form of a pulsed laser to measure ranges" (Husain 33). The sensors and software are "designed to detect and predict the behavior...of all road users," including cyclists and pedestrians (Waymo). The vehicle is expected to know its exact proximity to any object at any time, the speed limit, where it is, and where it's going.

Figure A.9. This is Jessamyn's findings section using MLA-style citations. Formal reports often rely on a blend of quantitative and qualitative research.

Feasibility of Driverless Vehicles 6

Findings

Cost of Autonomous Vehicles

Today, the cost of driverless vehicles is significant, ranging from $225,000 to $1,000,000. The software and hardware packages are the costly portion, ranging from $70,000 to $150,000 (Lienert). According to Kevin Clark, the CEO of Delphi Technologies, "the cost of that autonomous driving stack by 2025 will come down to about $5,000 because of technology developments and (higher) volume" (qtd. in Lienert). The cost of LIDAR is already declining. Velodyne, the primary producer of this technology, indicates that it will cut the $8000 cost of its most popular sensor in half. Given that one vehicle likely has more than one sensor, the cost reduction is significant (Davies). The industry is aware that the high prices are a significant barrier to advancement (Davies).

Maintenance

Many of the vehicles are current models that have been modified, such as the Toyota Prius, Lexus RX450H, and Chrysler Pacifica. The sensors and parts to make it autonomous are developed by and purchased from various tech companies. With a modified vehicle, the parts and repairs for the vehicle are similar to what we see now, with the notable exception of the hardware and software specific to the driverless package. Other

The Findings section is where most of your research lives. The evidence you collect should clearly define the issue.

This parenthetical citation shows how to cite an indirect source, which means that someone other than the author is cited.

Figure A.10. Jessamyn's report addresses ethical concerns regarding her topic.

Notice how this sentence uses parallel structure to list actions: ride, idle, (do not) maintain, and park. The sentence that follows stands out because it is short and direct.

An ethical presentation of data will not ignore potential problems that might result from the proposed solution.

Feasibility of Driverless Vehicles 7

vehicles that are developed from start to finish primarily by the tech companies will be more unique, and access to parts will be more limited. The driverless vehicles are expected to save maintenance costs in some ways just by removing the driver. Humans ride their brakes, idle unnecessarily, do not maintain a consistent (or often legal) speed, and park too close to their neighbor. AVs avoid these issues.

Safety

As previously mentioned, AVs are expected to be safer than current vehicles. Humans are prone to distraction and overreaction. They don't react when they should or, in theory, as quickly as a computer. AVs are programmed to follow the rules of the road. They do not speed up when a light turns yellow, cut people off in traffic to make a turn they almost missed, forget to use their blinkers, text while driving, drive under the influence, or suffer from road rage. When all of the reasons accidents happen are considered, it is easy to understand why AVs could reduce accident statistics and save lives.

Concerns and Controversies

Many people, customers and drivers alike, fear the reliance on software. Many companies have had similar concerns that their technology may not yet have "learned" enough to be self-sufficient. They've chosen to

Figure A.11. Jessamyn's report includes alternate points of view and credible research.

Feasibility of Driverless Vehicles 8

keep a driver in the vehicle to take over in an emergency and the

technology requires this (level three or four). This idea to keep the human

in the driver's seat misses the main reason for wanting to remove them in

the first place. Tesla, one of the companies that has chosen to form their

own technology in lieu of LIDAR, uses a system they've named "Autopilot."

In May 2016, a Tesla sedan ran into the broad side of a tractor trailer while

the Autopilot was engaged. "Tesla said that Autopilot didn't register the

white side of the trailer against the bright sky" (qtd. in Simonite). Neither

Autopilot nor the driver braked.

Other companies, like Waymo (Google's company) and NAVYA,

have been operating their vehicles without a driver for some time now and

have advanced their technology significantly. The cars have been in

accidents, though few, and nearly all were the result of cars with drivers

who didn't follow the rules (*On the Road*). As these vehicles become more

commonplace, which will be several years yet, these types of accidents

should rarely occur.

Market Analysis

Pew Research Center recently surveyed over 4,000 Americans to

determine their comfort and concerns with advances in technology,

including driverless cars and technologies that would replace human

This sentence shows how you can insert your own voice into formal report without using the first person.

Credible research is important. Pew Research Center is well-respected. The sample size (4,000 Americans) is large enough to provide meaningful results.

Figure A.12. Jessamyn cites her sources and includes a variety of sources in her report using MLA style.

This parenthetical citation shows how to cite a source that has two authors.

Surveys can be an effective way to quantify subjective experiences.

The parenthetical citation shows that this information comes from another source cited in the article by Smith and Anderson.

Feasibility of Driverless Vehicles 9

workers (Smith and Anderson). They learned that "Americans are roughly twice as likely to express worry (72%) than enthusiasm (33%) about a future in which robots and computers are capable of doing many jobs that are currently done by humans" (Smith and Anderson). The majority believe that such advances will create a larger income inequality and that "the economy will not create many new, better-paying jobs for humans if this scenario becomes a reality." With specific regard to autonomous vehicles, 39% expect the roads will be safer while 30% think the vehicles will make them less safe. In fact, 56% indicated that they would not ride in a driverless car (qtd. in Smith and Anderson).

Regulations

The regulations regarding automated driving systems (ADS) are not prohibitive. The NHTSA has established guidelines for states to help them safely integrate these vehicles onto their roadways. Most states require an application process for test vehicles, as well as licensing and registration that include more detailed information on vehicle ownership.

Regulations will likely adjust to state and federal levels as the technology progresses and as our government begins to realize that auto and tech companies looking at their bottom lines may not always be relied upon to move forward with caution or with the public's safety in mind. There

Figure A.13. Jessamyn's report includes complex data and uses tables to make the information easier to understand.

Feasibility of Driverless Vehicles 10

are suspicions that some have moved forward with pilot programs even when they knew of flaws in their systems and that a lack of rules is to blame (Wolverton). The NHTSA wants this technology to move forward, as many do. Yet it is hard to regulate what is not yet understood, and, in this case, it is hard to fully understand it until tested under true conditions.

Cost Comparison

If Tomorrow's Taxi Company were to replace all vehicles, the numbers would look tentatively like the following:

Table 1
Estimated Revenue for an Autonomous Vehicle Fleet

	Driver/Operator			Fuel/Energy	Maintenance	Insurance	Revenue	Net
	No. Cars	No.	Annual Cost					
Current Fleet	40	65	2,710,000	237,104	50,536	151,609	6,551,020	$3,401,771
AV Fleet 20XX	38	10	360,000	94,605	40,858	163,435	7,062,020	$6,403,122
AV Fleet 20XX	38	8	380,000	94,605	34,049	149,815	7,062,020	$6,503,551

This would be in addition to the purchase price of a complete fleet of driverless cars, which today would be very costly, approximately $9,400,000 for a fleet equaling the current size, plus the monitoring equipment.

Driver/Operator. Numbers include annual salary, benefits, and worker's compensation insurance for the drivers. Operators will replace the drivers, taking on the role of dispatching and monitoring the fleet remotely.

The formal report does not avoid complexity. The entire purpose of a formal report is to analyze a topic in depth.

Provide a title for all tables and visuals.

Information is fictionalized for this table. All costs, unless otherwise cited, are also fictional. Your research, on the other hand, should be real.

Figure A.14. Jessamyn's report uses bullets and numbers to emphasize important information.

Lead into a list with a complete sentence followed by a colon.

Use numbers and bullets to emphasize important information and make it more visually appealing.

Feasibility of Driverless Vehicles 11

Fuel/Energy: Tomorrow's Taxi Company's current vehicles are hybrids that average 50 MPG, driving approximately 2,339,000 miles per year, at an average of $2.90 per gallon. The AV fleet is electric and will get the equivalent of 93 MPG at a rate of $3.48 per 100 miles. The electric fleet should be able to drive approximately 2,522,150 miles per year.

Maintenance: Numbers for the current fleet are actual. Numbers for the AV fleet assume 3% of the IRS mileage reimbursement rate. This number reduces to 2.5% in 2021 with the assumption that the cost for parts, which are now quite expensive, will drop as mentioned. AVs are also expected to experience less wear and tear in most areas. This assumes the following:

1) Human error causes most accidents that result in repair.

2) Inconsistent driving practices, such as braking patterns, unnecessary idling, and not maintaining speed will be reduced in an AV.

What isn't clear, is the cost for parts and ease of replacement. While they are likely to spend less time in the service bay, they could be costly once there.

Insurance: This number is unknown for the AV fleet. Here it is estimated at the same rate of 12% of the IRS mileage reimbursement rate

Figure A.15. The discussion section is where Jessamyn will put the majority of her research for the formal report.

Feasibility of Driverless Vehicles 12

as it would be for a traditional vehicle. The numbers are expected to be lower as the technology becomes more established.

Revenue: These numbers are based on a rate of $2.80 per mile for both the current and AV fleet.

Net: This number is related to the direct cost of operating the vehicles and does not include numbers for any other operating costs for the company.

Discussion

The largest savings is in operational labor. The company would need to maintain a dispatcher and a small staff to monitor the position and condition of the vehicles. By removing the drivers, the company not only

> The discussion section is where you draw conclusions from your research and connect it to the issue you identified.

Feasibility of Driverless Vehicles 13

The change in labor structure also enables the company to keep the taxis on the road more often. The AV's eliminate downtime from breaks, sickness, and the general inattention humans have that software doesn't. The estimated annual amount of lost revenue from driver-related issues is approximately $30,600. According to Deloitte, one of the country's leading legal and financial consulting firms, they have witnesses changing trends in the automotive industry, an expansion they call the "mobility ecosystem" ("Future of Mobility Trends"). Deloitte states the following:

> Across the ecosystem, from auto retailers to insurance to finance, businesses are watching automotive industry trends and realigning to remain competitive and viable as the future of mobility unfolds. Companies that are preparing now—deciding where to play, transforming operations, implementing new technology, refocusing talent and marketing—will be well positioned to win. ("Future of Mobility Trends")

> This signal phrase attributes this information to Deloitte (a consulting firm) and ends with a parenthetical citation that includes the article's title that links to the entry on the works-cited page.

> Direct quotes that are more than four lines of text should be formatted as a block quote without quotation marks.

Figure A.16. Jessamyn provides recommendations that include several options.

Feasibility of Driverless Vehicles 15

Company would have the option to negotiate a purchase for the vehicles at a discounted rate.

Option Two – Test Vehicles

Under this scenario, Tomorrow's Taxi Company would lease-to-own two driverless vehicles from "Corporation A" at a rate of $20,000 per month for a period of six months. For the first six months, the corporation will

Feasibility of Driverless Vehicles 14

Recommendations

Given the relatively small size of the company, and its financial limitations in comparison to industry leaders, Tomorrow's Taxi Company would need to tread carefully into the area of driverless vehicles. The autonomous projects that require a driver's presence have been excluded from the discussion for safety and cost reasons. The following options are the most feasible at this time:

Option One – Partnership and Pilot

"Corporation X" seems to be the U.S. automotive company most eager to release a fleet of AV's. They've developed partnerships with a few other companies in order to pilot their vehicles and are interested in using Tomorrow's Taxi Company as well. For the first year, they would provide the thirty-eight vehicles. Tomorrow's Taxi Company would be responsible for maintenance and energy costs, insurance, and monitoring staff. They would oversee any issues with the vehicles, make changes and upgrades, provide parts, and train our staff. They would receive .5% of fare profits for this period for the use of the vehicles. At the end of the contract period, it could be renegotiated for another year if the technology is still needing improvement (which would likely be the case) or Tomorrow's Taxi

Recommendations typically come before the conclusion in a formal report.

A feasibility report ends with an analysis of options.

Not all formal reports will provide multiple options. Sometimes, a single option emerges as the best choice.

Figure A.17. This is Jessamyn's conclusion for her formal report.

Feasibility of Driverless Vehicles 16

Conclusion

The company will need to be prepared to make the transition to fully autonomous vehicles, likely within the next five years. The companies currently entering this field are large players in the automotive or technology industries, with the knowledge and financial resources to venture into this unchartered territory. Driverless vehicles are still a work in progress, though the progress is impressive. They are a new and exciting technology but also a frightening one for many. For that reason, they are under a microscope with various authorities and the community in general, and a company like Tomorrow's Taxi Company may be highly vulnerable should one of the vehicles have a safety issue. However, the taxi industry will be the first to integrate AVs regardless of such concerns, so it is important that the company move forward proactively but cautiously.

The concerns with the technology over safety, regulatory changes, and cost are there; however, at the current pace, these will soon be of little issue. The AVs, though not without their accidents, are still safer than Tomorrow's Taxi Company's current vehicles. As the numbers are tracked, studies are done, and more people experience the technology, a company with a driverless fleet will find itself well-positioned to succeed.

Conclusions usually follow the recommendation section in a formal report.

Conclusions usually review the benefits of your recommendation.

Conclusions often end with a forward-thinking call to action.

Figure A.18. Jessamyn's report ends with an MLA-style Works Cited page for her research.

Feasibility of Driverless Vehicles 18

Simonite, Tom. "Fatal Tesla Autopilot Crash is a Reminder that Autonomous Cars Will Sometimes Screw Up." *MIT Technology Review*, 30 June 2016,

---

Feasibility of Driverless Vehicles 17

Works Cited

Davies, Chris. "A Key Part of Many Autonomous Cars Just Got a Huge Price Cut." *SlashGear*, 2 Jan. 2018, www.slashgear.com/ velodyne-lidar-puck-autonomous-car-more-affordable-02513340/.

"Future of Mobility Trends." *Deloitte*: www2.deloitte.com/us/en/pages/ consulting/solutions/future-of-mobility-trends-industry-ecosystem.html. Accessed 3 Sept. 2020.

Husain, Amir. *The Sentient Machine: The Coming Age of Artificial Intelligence*. Scribner, 2017.

Lienert, Paul. "Cost of Driverless Vehicles to Drop Dramatically: Delphi CEO." *Insurance Journal*, 5 Dec. 2017, www.insurancejournal.com/ news/national/2017/12/05/473134.html.

National Highway Traffic and Safety Administration. "Automated Vehicles for Safety," U.S. Department of Transportation, www.nhtsa.gov/ technology-innovation/automated-vehicles-safety. Accessed 3 Sept. 2020.

On the Road to Fully Self-Driving. Waymo, February 2021. waymo.com/safety/.

This entry shows the standard format for an electronic source:

Last Name, First Name. "Article Title." *Site Name*, Day Month Year Published. URL.

This entry shows the standard format for a book:

Last Name, First Name. *Title: Subtitle If Any*. Publisher, Date.

For government documents, use the more specific agency name as the author. List the parent agency as the publisher before the URL.

Figure A.19. Figures and other additional material are often included in an appendix.

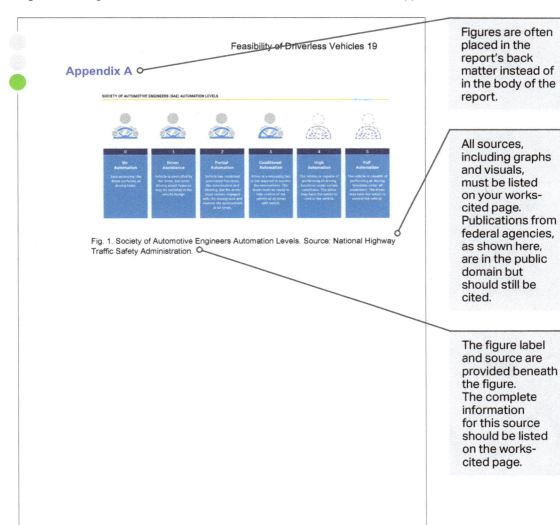

Feasibility of Driverless Vehicles 19

Appendix A

SOCIETY OF AUTOMOTIVE ENGINEERS (SAE) AUTOMATION LEVELS

Fig. 1. Society of Automotive Engineers Automation Levels. Source: National Highway Traffic Safety Administration.

Figures are often placed in the report's back matter instead of in the body of the report.

All sources, including graphs and visuals, must be listed on your works-cited page. Publications from federal agencies, as shown here, are in the public domain but should still be cited.

The figure label and source are provided beneath the figure. The complete information for this source should be listed on the works-cited page.

Glossary Index

The glossary index increases the usability of this book by providing definitions, chapter numbers, and page numbers where extended discussion of important terms can be found. All defined terms in the glossary can be found in bold within the text. Additionally, the bold numbers tell you where to find the definition of that term within the text. Additional page numbers direct you to where the term is discussed in depth.

The index does not include every mention of a word. If we did that, the entry for technical communication would be almost as long as the book itself. Instead, we have selected the most relevant locations for these terms. Additional page numbers refer to illustrations, models, examples, and extended discussion of these relevant terms. "See also" directs you to a similar term within the index. "See" directs you to the full entry for that term.

A

abstract: 307; model of, 318; *see also* summary.
A brief summary of an entire report, including the problem, methodology, results, and conclusions.

abstract language: 22
Words that describe ideas, feelings, and concepts that are understood through the mind rather than the senses; as opposed to concrete language.

accuracy: 23
The use of error-free information.

agenda: 187
An outline of the planned contents of a business meeting.

alignment: 66; 159, *see also* center alignment, justified alignment, left alignment, *and* right alignment.
The horizontal or vertical connection between separate elements that fall along straight lines.

alternate steps: 245
A type of instruction that allows for two or more ways to accomplish the same task.

ambiguity: 37
The presentation of two meanings at the same time that can lead to uncertainty.

analysis: 120, 294; *see also* causal analysis, comparative analysis, *and* feasibility analysis.
The act of examining an object or idea closely to understand its parts and how they work.

anchoring: 256; *see also* research.
A cognitive bias that favors the first piece of information you hear.

APA style: 127; 48, model of, 316–333
A style of writing and documentation defined by the American Psychological Association for behavioral and social sciences.

appendix: 278, 314
A section at the end of a report that provides additional information on subjects requiring further explanation or reference.

audience: 5; 200, 213, 250 *see also* user.
The group of people to whom a document or deliverable is addressed.

B

bar graph: 80
A visual representation of data in horizontal or vertical bars.

bibliography: 314; *see also* Chicago style.
A list of sources used for research in a report.

C

causal analysis : 296
A type of report that investigates why something happens.

causation: 297
The process of showing the direct relationship between a cause and effect; see also correlation.

center alignment: 66
A consistent connection in vertical text along a straight line at the center of a document, creating ragged left and right margins.

Chain Method: 226; *see also* Fork Method *and* Known-New Contract.
A way of explaining new information that begins with a topic sentence followed by a comment; subsequent sentences (chains) present new comments related to the topic.

characteristics: 218; *see also* class *and* name.
The unique set of traits that makes a term stand out within its class.

chart: 79; *see also* Gantt chart.
The organization of data to show relationships by using shapes, arrows, lines, and other design elements.

Chicago style: 128; 49
A style of writing and documentation used within the publishing industry and some business and history courses.

chronological organization: 74
A method of arranging information based on the progress of time.

circumlocution: 40
The use of evasive or excessive language to disguise one's meaning.

citation: 45
The academic convention that gives credit to the owner or creator of content and provides an ethical way to use content, sometimes referred to as a parenthetical or in-text citation.

claim: 140
The assertion of a particular message that requires evidence to establish its validity.

clarity: 15
A quality of writing that make it understandable and easy to follow.

class: 217; *see also* characteristic *and* name.
The category in which a defined word belongs.

cliché: 166
A phrase that is overused and shows a lack of originality.

clients: 12
A person or organization who pays for work from a technical communicator.

code of ethics: *see* ethics.

collaboration: 202
An activity or project where tasks are divided among several individuals who combine their separate efforts into a final product.

color: 65
A visual element that results from the perception of light on the surface of an object.

comparative analysis: 295

A type of report that identifies criteria for determining the best choice from multiple options.

conciseness: 20

The use of the right amount of detail and information.

conclusion: 312

The section of a report that ties the results together and often includes steps for implementing a solution or other recommendations.

concrete language: 222

Words that describe specific and tangible objects that exist in the world and are understood through the body's five senses; as opposed to abstract language.

context: 211

The setting or situation that gives words additional meaning.

contrast: 70

A way to create emphasis by highlighting differences between elements in a document.

copyright law: 131; 45

The legal protection that exists for people who own their content.

correlation: 297

The process of showing the approximate connection between two related factors; see also causation.

Creative Commons: 46

An organization that provides legal designations called licenses that allow copyrighted content to be used in an "open-source" style.

D

data: 23

Units of information used to create meaning.

definition: 211; *see also* extended definition, parenthetical definition, *and* sentence definition.

A statement that expresses the meaning of a word or group of words.

demographics: 14

The group characteristics of an audience.

description: 210; *see also* process description *and* product description.

A statement of the physical characteristics of an object.

design: 56; 243, 253

The intentional and planned presentation of information to an audience.

diagram: 80

A visual representation of how objects or parts of an object fit together.

digital literacy: 94

A form of technological knowledge that provides essential skills for finding, using, and sharing information.

directive: 185

A type of memo that issues an order to staff.

document design: 59; *see also* layout.

The deliberate organization of text and images on a page.

E

end matter: 314

The parts of a formal report that follow the main body, which may include a bibliography, appendix, glossary, and/or index.

end user: *see* user.

ethics: 31
An system of principles or morals that determine the actions of an individual or group.

etiquette: **172**; 184, 205
A code of behavior that explains the proper way to act in situations.

euphemism: 38
A word or set of words that replace other words to blunt or soften the intended message.

exploded diagram: 247
An illustration that allows the user to see all the parts and how they fit together.

extended definition: 219
An explanation over multiple sentences.

external proposal: 272
A persuasive form of business writing intended for audiences outside an organization.

F

fact: 23
Verifiable information.

fair use: 132
A narrowly defined legal use of copyrighted content for specific purposes.

feasibility analysis: 297
A type of report that helps determine if a strategy, plan, or design is a good idea based on finances, outcomes, or possibility.

feasibility report: 283; *see also* Appendix A.
A technical document that examines whether a proposed idea or product is likely to succeed.

fixed-order steps: 245
A type of instructions that must be performed in the order they are presented.

font: 63
The style of a set of characters used in typesetting and design.

Fork Method: **225**; *see also* Chain Method *and* Known-New Contract.
A way of explaining new information that begins with a topic and is followed by a sentence that starts with a word or phrase related to the topic that provides more information and furthers understanding about the topic.

formal report: 292
A long document that presents solutions to complex issues involving research, analysis, methodology, results, and recommendations.

format: 192
The style of a document based on how the sections and paragraphs are or are not indented; includes block, indented, or modified block formats.

freelancer/contractor: 342
A self-employed individual who performs duties for a defined period of time according to a contract agreed on with a client.

front matter: 306
The opening pages of a report that introduces the report's topic and often includes a letter of transmittal, title page, table of contents, and abstract.

G

Gantt chart: 276; *see also* chart.

A visual that represents tasks and timelines in collaborative projects where multiple assignments need to be completed at different times to achieve a specific outcome.

glossary: 217, 314

A list of terms used in a document, located near the end.

graph: 80; *see also* bar graph *and* line graph.

A visual that represents data points and allows the viewer to compare two or more variables.

grid: 68

The underlying structure used to help create visual order in documents.

group work: 202

A type of project where multiple individuals are given a task or set of tasks to complete regardless of individual strengths and skills.

H

heading: 72; *see also* subheading.

A word or phrase used as a title for a section of text.

honesty: 35

The state or quality of sticking to the facts and speaking the truth; one of the six ethical principles embraced by the society for technical communicators; *see also* ethics.

hypothesis: 109

An educated guess that serves as a testing ground for a research process.

I

idiom: 42

A phrase that has a specific meaning in one language that cannot be directly translated into another.

IEEE style: 130; 49

A style of writing and documentation defined by the institute for electrical and electronics engineers for use in the electrical, electronics, computer science, and computer programming fields.

illustration: 80

A visual representation of a physical object or concept, usually a sketch or drawing by hand or computer.

index: 314

A list of specific terms in a report and the page numbers where they can be found.

inference: 23

A conclusion that can be reached based on available data.

instructions: 234; 260

A set of detailed steps that function as directions for how to complete actions successfully; as opposed to procedures.

intellectual property: 131; *see also* copyright.

A product of the human imagination that is protected by copyright law.

internal proposal: 271

A persuasive form of business writing intended for audiences inside an organization.

introduction: 310

A section of a report that describes the problem and defines the report's purpose, scope, background, and method of analysis.

J

jargon: 43, 214
Words and phrases, technical or otherwise, that are unfamiliar to a general audience.

judgment: 24
The use of opinion-based, reasonable actions based on available facts, inferences, and values.

justified alignment: 67
A consistent connection in vertical text along a straight line at the left and right margins.

K

keyword: 163
A specific word or phrase that relates directly to a specific topic.

Known-New Contract: 225; *see also* Fork Method *and* Chain Method.
The principle of communication where the creator begins with what their audience knows before introducing new information.

L

layout: 56
The intentional placement of visual elements within a document

left alignment: 66
A consistent connection in vertical text along a straight line at the left margin, creating a ragged right margin.

letter of transmittal: 306; model of, 308
A brief document that precedes a report, informing the audience of the report's delivery.

levels of evidence: 303
A continuum on which each piece of evidence falls, with specialized knowledge at one end and general knowledge at the other end.

line graph: 80
A visual representation of data as points that are connected by a through line.

M

medium/media: 9, 90; *see also* multimedia communication.
A means of transmission to store or deliver information

memo: 185
A document distributed or displayed internally within a workplace to inform or remind the audience of a decision.

message: 6
The content of a document or deliverable.

methodology: 310
The section of a formal report that explains how research was conducted.

minutes: 187
A document that records the outcomes of a business meeting.

MLA style: 125; 48
A style of writing and documentation defined by the Modern Language Association for use in literature, languages, and the humanities.

mode: 89; 9, *see also* multimodal communication.
A style or manner in which communication occurs or is experienced, expressed, or done.

multimedia communication: 11, 89
A process involving the use of more than one means of information transmission.

multimodal communication: 10, 89
A process involving the use of more than one method of generating content.

N

name: 217; *see also* characteristic *and* class.
The specific word for a term, thing, or concept.

negation: 221
An explanation of what a term does not mean, used to help clarify meaning.

nested steps: 245
A type of instructions for a complex process that is broken down into substeps.

network: 352
The act of maintaining mutually beneficial professional relationships.

O

objectivity: 303
The use of external evidence and verifiable facts to support a conclusion.

P

page design: *see* document design, layout.

parallelism: 73; 241
A technique where elements in a group or list all take the same form, creating a pattern.

paraphrase: 123
A restatement of someone else's information in your own words.

parenthetical definition: 215
An explanation of a term immediately after its first use, typically enclosed in parentheses.

peer review: 119
A process of critical examination by a panel of experts used by scholarly publications before an article or other researched work is published.

plagiarism: 130
A violation of the scholarly and professional expectation that, unless otherwise cited, the ideas, words, and research you present in a document are your own.

precision: 20
A quality of writing that uses exact and specific language.

primary research: 117; 301
The process of acquiring new data that has not been collected before; as opposed to secondary research.

Problem-Solution Framework: 5; 267, 279, 293
An illustration of how technical communication moves from problem to solution by considering the needs of audience, purpose, and message.

problem-solution organization: 77
A two-part structure that describes an issue and suggests a response or resolution.

procedure: 236
An overview of the best methods required to complete a process; as opposed to instructions.

process description: 223
A document that explains how complex events occur with attention to sequence, timing, movement, and necessary tools.

product description: 222
A document that details an object's physical characteristics and ask questions that focus on elements that you can see or touch.

proposal: 264; *see also* solicited proposal *and* unsolicited proposal.

A form of business writing that persuades someone to approve a service or course of action.

proximity: 72

The spatial distance between two or more elements in a document that can create meaning.

public domain: 132; 46

The designation for content that is not under any form of legal copyright.

purpose: 5, 299

The reason behind a report's creation.

R

reference: 167

People who can talk about your work skills and experiences.

report body: 312

The sections of a report that contain the collected data, analyses, or results.

request for proposal (RFP): 270

A document companies use to gather information about services, not products.

research: 117; *see also* primary research *and* secondary research.

The act of looking for, collecting, and evaluating information on a specific topic.

research report: 281

A short document that presents data about a specific topic.

results: 312

A section of a formal report and part of the conclusion that presents the research outcomes.

rhetoric: 93

The artful use of language to persuade an audience.

rhetorical awareness: 93

The act of thinking critically about the choices involved in the creation of a persuasive appeal.

right alignment: 66

A consistent connection in vertical text along a straight line at the right margin, creating a ragged left margin.

S

sales report: 282

A short document that presents data about product sales over a specific period of time.

sans serif font: 64

Typography that is simplistic and uses more separation between characters to provide a cleaner appearance.

scanning: 24

The act of looking for a specific piece of information in a document; as opposed to skimming.

scope: 299

The boundaries of what a report will and will not include.

secondary research: 118; 301

The collection of existing data extracted from interpretations by other creators; as opposed to primary research.

sentence definition: 216

An explanation of a term within a complete sentence.

tone: 197
> The attitude conveyed by one's choice of words.

U

unsolicited proposal: 272
> A persuasive form of business writing that has not been specifically requested by its audience.

usability testing: 254
> An unbiased examination of the effectiveness of a document.

user profile: 14
> A collection of information about potential audiences.

user: 12; *see also* audience.
> The audience for a given technical communication.

V

variable-order steps: 245
> A type of instructions that allow a user to perform the suggested steps in any order.

W

white space: 70
> Any space surrounding figures, tables, visuals, or text that is otherwise empty.

work for hire: 132
> A type of contract where a hired creator releases ownership of the content they create to a client.

workplace communication: 178
> The exchange of information that takes place between individuals or groups trying to complete a job or task.

www.ingramcontent.com/pod-product-compliance
Lightning Source LLC
Chambersburg PA
CBHW080608060326
40690CB00021B/4621